Texts in Philosophy
Volume 20

History and Philosophy of the Life Sciences in the South Cone

Volume 10
The Socratic Tradition. Questioning as Philosophy and as Method
Matti Sintonen, ed.

Volume 11
PhiMSAMP. Philosophy of Mathematics: Sociological Aspects and Mathematical Practice
Benedikt Löwe and Thomas Müller, eds.

Volume 12
Philosophical Perspectives on Mathematical Practice
Bart Van Kerkhove, Jonas De Vuyst, and Jean Paul Van Bendegem, eds.

Volume 13
Beyond Description: Naturalism and Normativity
Marcin Miłkowski and Konrad Talmont-Kaminski, eds.

Volume 14
Corroborations and Criticisms. Forays with the Philosophy of Karl Popper
Ivor Grattan-Guinness

Volume 15
Knowledge, Value, Evolution.
Tomáš Hříbek and Juraj Hvorecký, eds.

Volume 16
Hao Wang. Logician and Philosopher
Charles Parsons and Montgomery Link, eds.

Volume 17
Mimesis: Metaphysics, Cognition, Pragmatics
Gregory Currie, Petr Koťátko, Martin Pokorný

Volume 18
Contemporary Problems of Epistemology in the Light of Phenomenology. Temporal Consciousness and the Limits of Formal Theories
Stathis Livadas

Volume 19
History and Philosophy of Physics in the South Cone
Roberto de Andrade Martins, Guillermo Boido, and Víctor Rodríguez, eds.

Volume 20
History and Philosophy of the Life Sciences in the South Cone
Pablo Lorenzano, Lilian Al-Chueyr Pereira Martins, and Anna Carolina K. P. Regner, eds.

Texts in Philosophy Series Editors
Vincent F. Hendriks   vincent@hum.ku.dk
John Symons   jsymons@utep.edu
Dov Gabbay   dov.gabbay@kcl.ac.uk

History and Philosophy of the Life Sciences in the South Cone

Edited by
Pablo Lorenzano,
Lilian Al-Chueyr Pereira Martins,
and
Anna Carolina K. P. Regner

© Individual author and College Publications 2013.
All rights reserved.

ISBN 978-1-84890-106-3

College Publications
Scientific Director: Dov Gabbay
Managing Director: Jane Spurr
Department of Computer Science
King's College London, Strand, London WC2R 2LS, UK

http://www.collegepublications.co.uk

Original cover design by orchid creative   www.orchidcreative.co.uk

Printed by Lightning Source, Milton Keynes, UK

All rights reserved. No part of this publication may be reproduced, stored in a retrieval system or transmitted in any form, or by any means, electronic, mechanical, photocopying, recording or otherwise without prior permission, in writing, from the publisher.

TABLE OF CONTENTS

Presentation XI
    Pablo Lorenzano
Introduction XIII
    Pablo Lorenzano, Lilian Al-Chueyr Pereira Martins, and
    Anna Carolina K. P. Regner
The Darwinian Concept of Causality 1
    Anna Carolina K. P. Regner
    1. The autonomy and nature of the causes: the ontological
       focus 2
       1.1. Causes as "means of modification" 4
       1.2. Causes as "power to produce" 5
       1.3. Causes as "agencies of modification" 7
       1.4. A view of Nature 9
    2. The explanatory condition: the epistemological dimension 12
       2.1. The search for the *vera causa* 15
       2.2. The universality of the causal relationship and of the
           concurrence of causes 16
       2.3. The causal structure 17
       2.4. Causes as necessary and / or sufficient "conditions" 18
       2.5. The "process-and-result" relationship as a causal
           condition 23
    3. Modes of causation: the confluence of the epistemological
       and ontological dimensions 24
       3.1. Teleological causality 27
       3.2. A view of Nature once again 36
       References 37
**Darwinism in the Mid-20th Century and the Relations between
Philosophy and Biology** 39
    Eduardo A. Musacchio
    1. Objectives of the present contribution 39
    2. Is it in conflict the notion of evolution in biology? 41
       2.1. The permanent aspects 41
       2.2. The nucleus of the theory of evolution and linked
           hypothesis 45
       2.3. Comparing theories of geologic change and fossil
           record 47
    3. Some difficulties in Darwinism (evident conflicts and
       proposed solutions during the mid 20th century) 54
       3.1. Continuous processes versus discontinuous processes 55

3.2. The notion of species within the framework of fossil record 57
   3.3. The sense of change and the progressive reduction of variability in the evolution 59
   3.4. The notion of increasing complexity 63
4. The philosophical substratum of currently accepted proposals of the theory of biological evolution and the conflictive parts 64
   4.1. The stable parts 65
   4.2. Focusing on the conflicts 67
5. Conclusions 74
   References 75

When (We) Biologists Get Close to Philosophy in Order to Make Our Work Better. Epistemological Proposal in Parasitology and Reflections on Its Use 79
Guillermo M. Denegri
1. A little bit of history 79
2. What parasitology is about and how a parasitologist works 80
3. Starting point of the epistemological problem 82
4. Methodology of research programmes applied to parasitology 84
5. How this proposal makes parasitology better 87
6. Final considerations 90
   References 91

Montparnasse Station. Vindicating Jacob's Reductionism in Functional Biology 95
Gustavo Caponi
1. Presentation 95
2. A few terminological explanations 97
3. Jacob's reductionism 103
4. Montparnasse Station 109
   References 111

About the Risks of a New Eugenics 117
Héctor Palma and Eduardo Wolovelsky
1. About the so-called "New Liberal Eugenics' 117
2. The eugenic movement 121
   2.1. Eugenic practices 125
   2.2. Medicalization: Eugenics as a technocratic practice 129
   2.3. "Racialism": Eugenics as an authoritarian practice 135
3. Can a New Eugenics be expected? 137

References 139
The Emergence of a Research Programme in Genetics 143
    Pablo Lorenzano
    1. Introduction 143
    2. Delimitation of the central problem 145
    2. Bateson's theory: "Mendelism" 147
    3. Bateson and the belief in the "promise" of Mendelism 150
    4. Mendelism: A research programme in genetics 154
    5. Conclusion 160
    References 160
On the Origin of «That Thing You Call "Species"» 171
    Santiago Ginnobili
    1. Introduction 171
    2. T-theoreticity 173
    3. Restating the question 174
    4. "That thing you call 'species'" 174
    5. "Species" as a NS-non-theoretical term 179
    6. Conclusions 180
    References 181
Alternatives in the Status of Darwinian Theory in Relation to the Epistemological Approach 183
    Gladys Martínez and Susana La Rocca
    1. Introduction 183
    2. Development 184
    3. Conclusion 193
    References 196
Chagas' Disease: History, Facts and Interpretations 199
    César Lorenzano
    1. Introduction 199
    2. Delaporte's central claims 200
    3. The structure of the disease 201
    4. The contributions of Carlos Chagas 202
    5. Romaña and the refoundation of the disease 205
    6. The chronic illness 207
    7. Chagas' disease, a heart disease 209
    8. The mistaken question 211
    9. Romaña and Mazza 213
    10. Carlos Chagas' article 214
    11. Mazza, the fraud 217
    12. Romaña's and Mazza's articles 219

13. The appearance of Romaña's sign as such...................................224
14. The justification of Romaña's sign...................................226
15. The estrangement 230
16. The scientific problem 232
17. The roles in history...238
    References..240

A Brief Reading of the Iatrochemistry in R. Bostocke: A Philosophical Summary of Man's / The Universe's 247
Ivoni Reis
    References..268

William Bateson's *Materials for the Study of Variation*: An Attack on Darwinism?..273
Lilian Al-Chueyr Pereira Martins
1. Introduction..273
2. Opinions on Bateson's position 274
3. What is Darwinism?..275
4. Darwin's original proposal 277
5. Darwin's research program 279
6. Bateson's early evolutionary work...................................280
7. Bateson's attitude in the Materials for the Study of Variation 282
8. Bateson's criticism of former methods 284
9. Discontinuous variations 286
10. Conclusion 291
    References..293

Newton Freire-Maia and Human Genetics in Brazil 297
Nadir Ferrari
1. Introduction..297
2. The research about genetic drift....................................298
3. Brief chronology 300
4. Scientist *en herbe* 300
5. Initiation ritual...303
6. At the UFPR..305
7. Conversion...306
8. Political views 308
9. Brazilian human genetics previous to Freire-Maia 310
10. Brazilian human genetics – generation one 312
11. Final comments..315
    References..316

The Beginning of Tropical Medicine in Argentina and Brazil........... 319

Sandra Caponi
    References 332

Presentation

The *Association of Philosophy and History of Science in the South Cone* (AFHIC) is a non-profit academic association, founded on May 5th, 2000, in Quilmes, Argentina, at the closing ceremony of the *2nd Meeting of Philosophy and History of Science in the South Cone.*

The creation of this Association was the result of the interest to deepen and strengthen the exchange between the researchers in Philosophy and History of Science from the countries of the South Cone, from the two first meetings that took place in Porto Alegre (Brazil, 1998) and Quilmes (Argentine, 2000) onwards. Since then, there have been biennial meetings organized as its responsibility.

The main aim of AFHIC is to contribute to a better understanding of science from a philosophical as well as a historical point of view in the Spanish- and Portuguese-speaking countries, especially in those which belong to the American South Cone, promoting a space for reflection, exchange, discussion, communication, and dissemination of such an understanding.

This volume is – with minor changes – the English version of *Ciências da Vida: Estudos Filosóficos e Históricos.* It is composed of refereed and, in some cases, opportunely modified contributions made by members of the *Association of Philosophy and History of Science in the South Cone,* some of them who are unfortunately no longer among us. It is an integral part of AFHIC's Book Series.

Its publication has been possible thanks to two men and their willingness to cooperate in this way with AFHIC's Book Series, and thus help us in achieving the main aim of our Association. Such men are the Editors of the Series "Philosophy", Vincent F. Hendricks and John Symons, to whom I wish to express my deep gratitude.

Pablo Lorenzano
Director of AFHIC's Book Series

Introduction

The present book – *Philosophy and History of the Life Sciences* – constitutes the second volume of the series *Philosophy and History of Science in the South Cone* and shows a significant sample of what the members of the Association of Philosophy and History of Science (AFHIC) devoted to these areas are investigating in the field of the Philosophy and History of the Life Sciences.

Starting with the contributions on Philosophy of Science, we have Anna Carolina K.P. Regner's article. It is on the leitmotif of the Darwinian explanatory task: the concept of causality. Then, Guillermo M. Denegri reflects on how philosophy can often contribute to make assumptions evident that pass unnoticed in the routine of scientific work, using as an example his work in experimental parasitology. Gustavo Caponi deals with the controversy about the relationship between physics and biology, defending a certain kind of reductionism of the molecular standpoint in functional biology, within François Jacob's adopted framework. Héctor Palma and Eduardo Wolovelsky analyze the premises present in the debate that signalizes the reappearance of eugenic programs called liberal eugenics, pointing out the differences in relation with those found in the eugenic programs of the end of the 19th Century and the beginning of the 20th Century. Pablo Lorenzano discusses the appearance of the first defined research program in genetics, Bateson's and collaborators' "Mendelism". Santiago Ginnobili analyzes the relationship between the theoretical term "species" and Darwin's theory of natural selection from a structuralist point of view. Susana La Rocca and Gladys Martínez deals with the diverse alternatives on the *status* of Darwinian theory from an epistemological point of view.

Going to the works on History of Biology and Medicine, César Lorenzano makes a critical analysis of Delaporte's version of the history of the Chagas' disease, giving a second glance to the contribution of Salvador Mazza. Ivoni Reis comments Rychard Bostocke's contributions to medical chemistry and its context. Lilian A.-C. Pereira Martins reviews the meaning of the term "Darwinism", discussing whether William Bateson's book *Materials for the Study of Variation* should be considered an attack to "Darwinism" as some authors do. Nadir Ferrari makes an examination of Newton Freire-Maia's contributions to human genetics in Brazil. Sandra Caponi analyzes the way in which Brazilian and Argentinean researchers around the end of the 19th Century and the beginning of the 20th Century build their research programs on tropical diseases.

The Philosophy and the History of the Life Sciences are interdependent metascientifc studies that often complemented each other, and they currently occupy an important place in the international scene. These studies offer a wide range of possible approaches. This diversity is reflected in the contributions made by researchers of the AFHIC that form this volume.

Finally, we would like to thank all those who, in one way or another, collaborated with this publication, including those who acted as anonymous referees of the works that are part of it.

The editors

The Darwinian Concept of Causality

Anna Carolina K. P. Regner
University of Vale do Rio dos Sinos (UNISINOS)

Causal inquiry is the leitmotif of the Darwinian explanatory task. Not only does Darwin aim at finding the *vera causa* of the origin or production of new species in Nature, but his legacy to future generations of investigators bears on a causal agenda:

> A grand and almost untrodden field of inquiry will be opened, on the causes and laws of variation, on correlation, on the effects of use and disuse, on the direct action of external conditions, and so forth. (Darwin 1875, p. 426)

In the *Origin*, Darwin never doubts that there is a causal legality to preside the relationships within Nature. Nevertheless, Darwin is not a professional epistemologist worried about previously defining the terms "cause" or "effect" he uses. Hence, in order to grasp the meaning of Darwinian causality it may be fruitful to look for "cause", "effect", and cognates in the text of the *Origin*, and analyze the concepts associated to these phrases. By following this procedure, we find two mutually related dimensions of causality in the Darwinian explanation: the ontological and the epistemological one. The first one points to the meaning of cause as a power of an agency or process acting in Nature. This meaning serves as a foundation for the epistemological dimension. The latter one establishes the conditions to be fulfilled by that agency or process in order to objectively (empirically) be an explanatory power. Both dimensions finally converge toward one another through the analysis of the "modes of causation" found in the *Origin*, making room for thinking about the possible role of a "teleological causality" in the Darwinian explanation.

1. The autonomy and nature of the causes: the ontological focus

Darwin's conception of causes as having an objective existence rooted in Nature can be apprehended from the autonomy he attributes to their existence and operation in relation to our knowledge, as well as from the intrinsic legality he attributes to Nature viewed as a self-regulated system. He admits that causes determine the course of phenomena, even if they are imperceptible, unknown or obscure to us:

> The causes which check the natural tendency of each species to increase are most obscure. (Darwin 1875, p. 53);

> We are profoundly ignorant of the cause of each light variation (Darwin 1875, p. 158);

> [...] we see much variability, caused, or at least excited, by changed conditions of life; but often in so obscure a manner, that we are tempted to consider the variations as spontaneous. (Darwin 1875, p. 410)

In several passages Darwin refers to the action of "unknown causes" (Darwin 1875, pp. 130, 207, 208, 257), "unknown favorable conditions" (Darwin 1875, p. 126), "causes that can be hardly seen" (Darwin 1875, p. 171), or "non perceptible by us" (Darwin 1875, p. 7). At least part of the difficulty in seeing causes is due to limitations of our mental habits (Darwin 1875, p. 422). But this difficulty may also reside in the very complexity of the cause (Darwin 1875, pp. 100-101). Darwin explicitly refers to our "ignorance of the causes" concerning central topics in the *Origin*, as those on variation (Darwin 1875, pp. 159, 174), cases of correlation (Darwin 1875, p. 228), rates of changes of species (Darwin 1875, p. 409), and cases of apparent difficulties (as in Darwin 1875, pp. 175, 179, 190).

However, the ignorance of the causes doesn't make our knowledge impossible. Rather, the appeal to our profound ignorance together with a detailed examination of the subject at hand becomes a main strategy used to preserve the nucleus of Darwin's theory from the most severe criticisms (Darwin 1875, pp. 160, 181, 249). Such limitations do not undermine the central points of the Darwinian theory, namely, the community of descent with modification and the conclusion that the structure of each living being is the result of its usefulness to its possessor, nor do they undermine the success of the Darwinian treatment of the apparently most important difficulties faced by his theory faced:

> [...] we are far too ignorant to speculate on the relative importance of the several known and unknown causes of variation; and I have made these remarks only to show that, if we are unable to account for the characteristic differences of our several domestic breeds, which nevertheless are generally admitted to have arisen through ordinary generation from one or a few parent-stocks, we ought not to lay too much stress on our ignorance of the precise cause of the slight analogous differences between true species. (Darwin 1875, p. 159)

> It is scarcely possible to decide how much allowance ought to be made for such causes of change, as the definite action of external conditions, so-called spontaneous variations, and the complex laws of growth; but with these important exceptions, we may conclude that the structure of every living creature either now is, or was formerly, of some direct or indirect use to its possessor. (Darwin 1875, p. 160)

> Finally, then, although we are as ignorant of the precise cause of the sterility of first crosses and of hybrids as we are why animals and plants removed from their natural conditions become sterile, yet the facts given in this chapter do not seem to me opposed to the belief that species aboriginally existed as varieties. (Darwin 1875, p. 263)

As objective entities, "causes" enjoy one of three referential "status": as "means" of modification / transformation, as a "power" to act, and as a "subject" endowed with this power. Fundamental passages of the *Origin* exemplify these conditions:

> [...] I am convinced that Natural Selection has been the most important, but not the exclusive, means of modification (Darwin 1875, p. 4).
>
> Over all these causes of Change, the accumulative action of Selection, whether applied methodically and quickly, or unconsciously and slowly but more efficiently, seems to have been the predominant Power. (Darwin 1875, p. 32)
>
> As man can produce, and certainly has produced, a great result by his methodical and unconscious means of selection, what may not natural selection effect? Man can act only on external and visible characters: Nature, if I may be allowed to personify the natural preservation or survival of the fittest, cares nothing for appearances, except in so far as they are useful to any being. She can act on every internal organ, on every shade of constitutional difference, on the whole machinery of life. Man selects only for his own good: Nature only for that of the being which she tends. (Darwin 1875, p. 65).

1.1. Causes as "means of modification"

In his *Historical Sketch*, Darwin explicitly excludes Buffon from the gallery of "evolutionists". Buffon would not have entered into the "causes or means of transformation of species" (Darwin 1875, p. xiii). Already in his *Introduction*, Darwin affirms that a conclusion favorable to the evolutionist view would be unsatisfactory, if it could not be shown how species were modified and co-adaptations acquired. In the following, Darwin considers of highest importance to reach a clear insight into the "means of modification and co-adaptation" (Darwin 1875, pp. 2-3). At the end of his *Introduction*, he affirms to be convinced that the "natural selection" has been the most important, although no exclusive, "means of modification" (Darwin 1875, p. 4). At the end of the first chapter, "natural selection" is pointed out as "the predominant power" (Darwin 1875, p. 32).

As the "means" through which changes are operated, "causes" hold the connotation of being "mechanisms" that describe or allow us to see *how* changes are produced. "Causes" make then possible our intelligible access to Nature's operations as long as these operations become an "object" to be known by us. Very often, the causal relationship is given as a "process" and, under such a connotation; "cause" is rather given in terms of a "power". In particular, in the exemplary situation of "natural selection", the operation of "cause" as a means through which certain results are obtained is equally seen as a "power" to act internal to the very course or process, acting through an internal chain of (light) modifications. There are several passages where Darwin literally refers to the "modification process" or to the "course" of natural selection

> Whatever the cause may be of each slight difference between the offspring and their parents – and a cause for each must exist – we have reason to believe that it is the steady accumulation of beneficial differences which has given rise to all the more important modifications of structure in relation to the habits of each species (Darwin 1875, p. 132).
>
> Any change in structure and function, which can be effected by small stages, is within the power of natural selection (Darwin 1875, p. 401).

It is under the perspective of a "process" that Darwin builds the bridge between the action of selection by man and natural selection, when he compares the effects of natural selection with those produced by the "unconscious selection" by man (Darwin 1875, p. 270), and when he compares the limitless amount of variation that can be operated through the slow and enduring power of the selection of Nature with the feeble results accomplished by man's selection (Darwin 1875, p. 65).

1.2. Causes as "power to produce"

In the text of the *Origin*, "cause" as a "power" to produce points to four fundamental connotations. "Cause" is that which: creates, originates, makes to exist (and makes to cease to exist); leads to a certain result; forms / develops something; preserves and accumulates something. In all these operations, the idea of "process" of production (or extinction) acting on what is "natural" or "regular" prevails. "Natural selection", as well as the state-of-things named "struggle for existence", on which the former is based, exemplifies the connotation of "cause" as a "power to produce". "Natural selection" causes much extinction and "leads to" divergence of characters (Darwin 1875, pp. 3, 103), through the preservation of the useful variations and the elimination of the injurious ones. Darwin says that such results "follow from" the "struggle for existence", "due to which" the preservation and extinction process take place. That is, "due to which" the Principle of Natural Selection as defined by Darwin (Darwin 1875, p. 49) takes place. The "struggle for existence", in turn, unavoidably follows from the high tax in that the organic beings tend to grow (Darwin 1875, p. 50).[1]

In the central passages of the *Origin,* the prevailing situations are those in which a cause "leads to" necessary or unavoidable effects or situations in which the necessity of a causal relationship is required as a parameter that drives the course of the investigation. For instance: each creature's improvement in relation to its life conditions "unavoidably leads" to a gradual progress in organization (Darwin 1875, p. 97); and the high advancement of whole classes or of certain members of each class does not "necessarily lead to" the extinction of those with whom the first do not enter in competition (Darwin 1875, p. 99). The need for a necessary causal-effect link seems to be reserved to the ingredients that constitute the "hard nucleus" of the Darwinian view of Nature and its Principle of Natural Selection. Outside of that nucleus, the direct action of life conditions leads to defined "or" indefinite results (Darwin 1875, p. 106), and the fact that beings inferior on the organic scale are more variable than the superior ones, "probably" follows from their lesser specialization (Darwin 1875, p. 131). The elimination of the sterility of the ancestor species "follows from" the same cause that allows the domestic animals to cross freely, and this cause "apparently" follows from the fact that domestic animals are gradually accustomed to frequent changes in their life conditions (Darwin 1875, p. 405). And the cause of this fact belongs to the cast of "ignored" or "obscure causes".

A large number of cases of the causal relationship seen as a "process" are under the connotation of "causing" as "forming / developing" referring to phenomena as distinct as the "formation" of geological strata and fossil layers, and the formation and development of biological entities on different levels, from embryos and rudimentary

---

[1] A particular causal relationship holds in the case of the "struggle for existence" and the Principle of Natural Selection, and exhibits a clear teleological dimension (Regner 2001).

organs to varieties, species, whole groups, genera, subfamilies, families, and classes. Such a formation process operates through successive steps and brings to our attention the connotation of "to cause" as "to preserve and to accumulate" modifications leading to the production of new forms. In relation to the selection by man, the key is the power of cumulative selection (Darwin 1875, pp. 3, 22, 64-65, 85). In relation to the power of selection by / within Nature, in several passages, from the very beginning of his work (Darwin 1875, p. 32), Darwin refers to the accumulative action of selection as the predominant power of change. Several meanings of "cause" become integrated under the connotation of cause as a "power to preserve and accumulate", as it can be found in the action of "natural selection": it is the "means of modification" through which that power "to originate" a new form from a former one is exerted as an internal condition to the process of their production. But in order to characterize a process of production of new forms in Nature by means of preservation and accumulation, which must happen "in a given direction", we must think of causality in terms of an *agency of the process*.

1.3. Causes as "agencies of modification"

When thinking about causes in terms of a "power to produce (or eliminate)", something leads us to think about causes in terms of the agency that exerts this power. The uniform action of the same cause produces the same effects:

> [...] if the same cause were to act uniformly during a long series of generations on many individuals, all probably would be modified in the same manner (Darwin 1875, p. 6).

And this cause is not thought of as an "external" agency to the course of the phenomena:

> [...] species are produced and exterminated by slowly acting and still existing causes, and not by miraculous acts of creation (Darwin 1875, p. 427).

"Agencies" that are still in action – even if ignored, unnoticed or imperceptible (Darwin 1875, pp. 295, 297, 346) – are used to explain past geological changes (Darwin 1875, pp. 75, 266). Rather than by their "visibility" to the senses, the explanation of the past by means of agencies still at work results from assuming the "uniformitarianism" and "gradualism" of "natural" and "ordinary" agents without appealing to "super-natural" forces. In this way, that which "causes" or "produces" reveals itself as an "autonomous" and "natural" power to act on what is given to our experience.

In several references that Darwin makes to *actions* that produce effects, "causes" are presented as "subjects" of these actions, and they may be individuals, like insects (Darwin 1875, pp. 77, 116), birds (Darwin 1875, pp. 346, 326, 345), or females that select males according to their beauty patterns (Darwin 1875, p. 70). But "causal agencies may also be state-of-affairs", "conditions of existence" (Darwin 1875, pp. 5, 22, 158, 106, 171, 262), habits (Darwin 1875, p. 22), modifications related to behavior patterns like mimicry (Darwin 1875, p. 377), the action of use / disuse of parts (Darwin 1875, pp. 400-401), and "laws".[2] Among all these agencies, there is also the human agent:

> [...] it is the will of man which accumulates the variations in certain directions; and it is this latter agency which answers to the survival of the fittest under nature (Darwin 1875, p. 108).

Man's selective action, whether intentional or not, produces new races (Darwin 1875, pp. 107-108, 269, 295, 292, 349). But it is in those passages related to "natural selection" that the causal power is more frequently put in terms of the exercise of a power that belongs to a "subject" who gives direction to the "accumulative" character of the change thus produced. And "natural selection" gives "direction" to the process as if it were a "subject" with a power to act greater than man (Darwin 1875, p. 65). It is also by bearing the *status* of such a powerful agent that "natural selection" becomes the only possible explanatory power against the competitive doctrine of "Creationism" and is supported by the new science of Geology:

> If species had been independently created, no explanation would have been possible of this kind of classification; but it is explained through inheritance and the complex action of natural selection, en-

---

[2] Considering the various agencies that concur to the production of the countless and complex co-adaptations of structures in Nature, Darwin is inclined to give less weight to the direct action of environmental conditions than to the tendency of organisms to vary due to causes that we ignore (Darwin 1875, p. 107).

tailing extinction and divergence of character, as we have seen illustrated in the diagram (Darwin 1875, p. 104).

Natural selection acts only by the preservation and accumulation of small inherited modifications, each profitable to the preserved being; and as modern geology has almost banished such views as the excavation of a great valley by a single diluvial wave, so will natural selection banish the belief of the continued creation of new organic beings, or of any great and sudden modification in their structure (Darwin 1875, pp. 75-76).

1.4. A view of Nature

The complex action of "natural selection" as a "natural" power brings to light a network of conditions that that shapes a certain view of Nature. Darwin offers two apparently conflicting conceptions of Nature: as an "object" to be determined by us and as a "subject" that acts in an autonomous way:

> [...] it is difficult to avoid personifying the word Nature; but I mean by Nature, only the aggregate action and product of many natural laws, and by laws the sequence of events as ascertained by us. (Darwin 1875, p. 63).

> Nature, if I may be allowed to personify the natural preservation or survival of the fittest, cares nothing for appearances, except in so far as they are useful to any being. She can act on every internal organ, on every shade of constitutional difference, on the whole machinery of life. Man selects only for his own good: Nature only for that of the being which she tends. (Darwin 1875, p. 65).

Although there is no room here to discuss that apparent conflict,[3] it is not difficult to infer from both "definitions" a view of Nature endowed with an intrinsic legality that determines the course of phenomena and underscores the multiple connotations of the Darwinian causality.

What kind of "legality" is that? Following closely the text of the *Origin*, we found that such an "order" is mainly "causal" and operates on different levels: as "principles", "laws", "rules", and "tendencies". Although these phrases may indicate different epistemic status, Dar-

---

[3] For a discussion of the conflict see Regner (2001).

win many times interchanges them and always refers to the production of new species as a "regular" and "natural" phenomenon, as opposed to a "casual" and "miraculous" event (Darwin 1875, pp. xx-xxi). All those instances of "order" exhibit a double dimension, both epistemic and ontological, in the *Origin*.

"Principles" establish norms and guidelines to be followed. Very often they appear in the text as causal agents, producing effects, or the the means by which effects are produced.[4] In some other passages, the combination of principles puts them in action, as in the case of the principle of benefit that tends to act in combination with the principle of natural selection and the principle of extinction (Darwin 1875, p. 90).

"Natural laws" are "secondary causes" (Darwin 1875, p. xvii). In keeping with what was common in his time, Darwin refers to "secondary causes" and "primary causes", and to the former ones as "immediate causes" (Darwin 1875, p. 212) or "proximate agencies" (Darwin 1875, p. 174). He refers to "primary causes" as those that act more intimately on the phenomena and present themselves as "ultimate causes" in the sense of being the last legitimate explanatory instance that can be reached (Darwin 1875, p. 393), or in the sense of being beyond our knowledge through "rules" or "secondary laws" (Darwin 1875, pp. 252 and 263).[5] "Laws" subject and regulate different kinds of phenomena. In certain circumstances, the cause is not the "law", but is inferred from the law (Darwin 1875, p. 248). In other passages, however, Darwin says that certain occurrences are "owed to" or "attributed to" certain laws (Darwin 1875, pp. 175, 177), explicitly referring to "laws" as "causes" (Darwin 1875, p. 299). On the one hand, Darwin underlines the uniformity and universality of the causal action of laws (as in Darwin 1875, pp. 89, 131, 362). On the other, he admits that "singular laws" determine, for instance, the reproduction in captivity (Darwin 1875, p. 7). Hence, there are different kinds and levels of "laws" and "causal laws".

---

[4] Diversified examples of principles as causal agents can be found in Darwin (1875, pp. 22, 23, 81, 87, 170).

[5] Contrary to what was common at his time, Darwin does not use "primary causes" or "ultimate cause" in the sense of the "Final Cause" that would complete the chain of causal conditions.

The "regular" and the "natural" equally mark those guidelines Darwin refers to as "tendencies". Many of them are endowed with causal force, or they are objects of causal action, or express a certain preferential causal behavior. The expression of the organic beings' causal behavior in terms of "tendencies" reflects two mutually related conditions. One condition has to do with the compound of interacting forces that indicates the "natural" or "normal" course to be followed and the result to be obtained according to the nature of the organic beings involved and the nature of their "conditions of life". The other condition consists of discriminating a "preferential state", among a range of possible states resulting from that interaction. As a general "norm", there is a tendency toward the "survival of the fittest", and it should be noted that this phrase encompasses a complex network of meanings.

Rather than expressing an absence of universality and necessity, "tending toward" expresses a mode-of-being of the phenomena and their reciprocal relationship in Nature. This mode is intrinsically dynamic, multi-factorial, and relational. Under the major determination of the Principle of Natural Selection that mode leads to the "survival of the fittest". As a central point to the Darwinian theory, the explicit reference to a "tendency to produce" (to bring into existence), or to a "tendency to eliminate", generates a "chain of tendencies":

> Dominant species belonging to the larger groups within each class tend to give birth to new and dominant forms; so that each large group tends to become still larger, and at the same time more divergent in character. (...) This tendency in the large groups to go on increasing in size and diverging in character, together with the inevitable contingency of much extinction, explains the arrangement of all the forms of life in groups subordinate to groups, all within a few great classes, which has prevailed throughout all time. This grand fact of the grouping of all organic beings under what is called the Natural System, is utterly inexplicable on the theory of creation (Darwin 1875, p. 413).

On the basis of the continuous tendency to diverge in characters one can understand another tendency, namely, the one according to which old and extinct forms fill out the gaps among the existent forms (Darwin 1875, p. 414). In any genus, the species that already quite differ amongst themselves "will usually tend to produce" the largest number of modified descendants (Darwin 1875, p. 93). During a long course of modification, the small differences will tend to be increased in larger differences (Darwin 1875, p. 413). There are several passages where Darwin refers to the tendency of the new varieties "to exterminate" their closest forms (Darwin 1875, p. 86), each new form tending to take the place and finally to exterminate its own parental and less improved form (Darwin 1875, p. 134). Related to the condition of "production" and "extermination", there are the tendencies of each species to increase in number inordinately (Darwin 1875, pp. 297, 413), of the dominant species to spread and supplant many others (Darwin 1875, p. 102), and of the species of the less vigorous groups tending to be extinguished without leaving modified descendants (Darwin 1875, p. 314).

In the Darwinian Nature, therefore, the natural course of events exhibits a "necessity" that gives a distinguished meaning to the "causality" that explains the order of Nature. This necessity is not imposed on facts such as a chain that linearly and forcefully drags them. Rather, this necessity is that of an order that feeds itself back from its own dynamism, and establishes a unifying thread to multiple factors without removing from each one of them and from the bunch of their relationships the richness of their several possibilities to come into being. Such a necessity does not reduce to a simplified pattern the possible interaction of several factors and forces that can be added or opposed to each other.

2. The explanatory condition: the epistemological dimension

The access to the understanding of that order of Nature is made possible by the epistemological dimension of the "causal relationship". Not "seeing" the causes may lead to invoke cataclysms or "invent" laws, and to attribute results to chance. But it does not necessarily have to be so. Considering the literal references that Darwin makes to what is "modified", "changed", "altered", "produced", "originated / created", "formed", or to what "suffers change", he extends his search for "causes" to the most different areas and objects of investigation, ranging from the formation of rocky scarps (Darwin 1875, p. 267) and sedimentary deposits (Darwin 1875, p. 277) to the production of "belief", and to the consideration of the effects of the theory of natural selection on the study of Natural History (Darwin 1875, p. 404); from the general production / creation of beings (Darwin 1875, p. 423) to the specificity of the production of ants casts and cells in the beehives (Darwin 1875, pp. 220 and 233). The focus of Darwin's investigation on what is "modified" or "produced" is expected to be on "species", but this focus literally reaches other levels of organic units, in their several aspects: genera, groups, classes, families, varieties, races, and individuals. In more generic references to what is modified, Darwin speaks of "life forms", adult / larval forms or "divergent forms" (Darwin 1875, p. 100); "new and improved" forms; "modified descendants", "progenitors" (Darwin 1875, pp. 320, 396), "colonists / immigrants" (Darwin 1875, pp. 354, 355); "productions of Nature" (Darwin 1875, pp. 161, 426), "of the world" (Darwin 1875, p. 314) – all of them encompassing different levels of specifications (Darwin 1875, pp. 340, 354, 400, 406, 415). Less inclusive biological units, like "embryos" (Darwin 1875, pp. 249, 390-391), "hybrids", "organic structures", "organs", "original patterns of structure" (Darwin 1875, p. 383), "reproductive system" (Darwin 1875, pp. 7, 259, 260), "fertile eggs / seeds" (Darwin 1875, pp. 387, 258), "rudimentary parts" (Darwin 1875, p. 400), "characters" (Darwin 1875, pp. 372, 379) – all can undergo modification, and become Darwin's target when he talks about the "power to produce".

The sort of things that are listed as "causes" in the *Origin* is diversified. "Causes" may be *fundamental properties*, such as variability; *principles*, such as the principle of inheritance or the principle of analogous variation; *laws*, such as the law of variation, correlation, or the general laws that govern the animal kingdom; the *history* of a certain organic form; or *phenomena* as varied as the cellular multiplication by division, the geological subsidence necessary for the accumulation of fossil deposits, of modification in different groups, migrations, etc.; *behavioral factors*, such as habits, that produce structure modifications of structure and mental characteristics (Darwin 1875, pp. 206, 211, 212, 416) and have some influence on acclimatization (Darwin 1875, p. 113); diverse *state-of-affairs* at different levels of complexity, from the generic reference to "physical conditions" (Darwin 1875, p. 415), "life conditions" (Darwin 1875, pp. 5, 31, 106, 107), "variability" (Darwin 1875, pp. 62, 260, 410), to specific situations, such as the occupation of the place of a former species by a new one, leading to the extermination of the first one (Darwin 1875, pp. 138, 296). Successions of state-of-affairs generate "causal chains" that constitute a central line in the argument of the *Origin*. Linking together all the factors to which Darwin refers to when he says that a certain modification or production is "due to", there is the case of a "process" being the sort of thing that is a "cause". This is the exemplary case we find in the nature of "natural selection" and in the causal role it plays.

Looking at what can be modified or produced, that is, to the "effects", we also find *properties*, *state-of-affairs*, *behavioral* and *mental* qualities or states (Darwin 1875, pp. 170, 209, 210, 211, 220). The "effects" to be focused range from the production of new organic forms to the acquisition of some physical and mental properties such as: "sterility" (Darwin 1875, pp. 235-236, 250, 256), the "revolving power by plants" (Darwin 1875, p. 197), the "clothing" exhibited by the imitating animals in the mimicry (Darwin 1875, pp. 376, 377), the "metamorphose stages developed by the insects" (Darwin 1875, pp. 386, 395), wide "geographical distribution", "gradations" (Darwin 1875, p. 168), as well as "variations" or "modifications in the direction requested by the theory" (Darwin 1875, p. 279), acquisition of language (Darwin 1875, p. 370), knowledge (Darwin 1875, pp. 162, 285, 345, 428), "habits" (Darwin 1875, pp. 165, 209), "instincts", "functions" (Darwin 1875, pp. 165, 168), "dispositions" (or even the natural indisposition to admit certain beliefs, and the smaller or larger mental flexibility to acquire new ways of looking at things – Darwin 1875, pp. 107, 123, 126, 134, 423, 426).

The causal attribution allows different epistemic degrees: sometimes "we can certainly attribute" a given cause to an effect (Darwin 1875, p. 208), attribute a given effect "mainly" to a certain cause (Darwin 1875, p. 344), "falsely attribute" a cause to an effect, or attribute a cause in terms of what the effect is "truly simply due to" (Darwin 1875, pp. 116-117). The appropriate causal attribution has to meet two key requirements: to determine *what* can contain the "conditions" for the given occurrence, and the nature of the relationship that can be established between "condition" and "conditioned". Tracking the use Darwin makes of "cause" in his text, we see that "causes", "influence" or "determine" what happens (Darwin 1875, pp. 115, 157, 169), but that such an influence or determination is not exerted by all "causes" in the same way. The degree of determination imposed by "causes" on "effects" depends on the weight and nature of the relationships of the "cause" we are looking at with other factors that can delay or favor its action and the production of new forms by "natural selection" (Darwin 1875, pp. 82-83). Physical changes condition the relationships among the organic beings:

> Species, however, probably change much more slowly, and within the same country only a few change at the same time. This slowness follows from all the inhabitants of the same country being already so well adapted to each other, that new places in the polity of nature do not occur until after long intervals, due to the occurrence of physical changes of some kind, or through the immigration of new forms (Darwin 1875, p. 270).

But theese are not the most important conditions. Darwin says it is very futile to look at changes of currents, climate or other physical conditions as the cause of the great mutations in the forms of life throughout the world, under the most diverse climates (Darwin 1875, p. 299). The nature of the organisms and the relationships among organic beings in the polity of Nature are by far much more important.

Darwin considers the access to "causes" provided by his theory as "fully declared" by the geological and paleontological evidence and as capable of promoting the advancement of the investigation in Natural History. The issue then to be discussed in what follows is: what are the conditions to be met in order to recognize something as a "cause" and relationships as being "causal" ones?

2.1. The search for the *vera causa*

Darwin clearly sets his explanatory goal in terms of the search for a *vera causa* for the production of new species in Nature. This goal articulates the ontological and epistemological dimensions of his view. Darwin finds strange the conclusion eminent naturalists arrived to in relation to the existence of the so called species belonging to each genus. Some of these species would be "real", but others would not be, and this would mean that the former ones would have been created independently. In contrast to that, "natural selection" as encompassing that which Darwin calls "ordinary generation" and "descent community" provides a *vera causa* for the production of new species.

The *vera causa* is a "natural" cause to be found within the scope of ordinary processes or agencies that take place in Nature and are able to explain phenomena apparently so diversified as the geographic distribution of organic forms and their similarities, as opposed to the appeal to miraculous intervention. Darwin says that those who reject the explanation by means of ordinary generation with subsequent modification have to invoke the agency of a miracle (Darwin 1875, p. 320). The community of descent is the *vera causa* capable of explaining similarities between the organic forms (Darwin 1875, p. 125). To admit Creationism would mean "to reject a real for an unreal or at least for an unknown cause" (Darwin 1875, p. 130). The *vera causa*, therefore, is a "real" cause found in the ordinary, regular course of Nature, and thus a cause that can be known by investigation and can explain the deepest bond among organisms within the order of Nature.

2.2. The universality of the causal relationship and of the concurrence of causes

In the Darwinian Nature, an efficient cause is always necessary for the production of the phenomena to be investigated, even if the causes are imperceptible or unknown. More than once Darwin says that whatever it may be, there must be a cause for each light variation (Darwin 1875, pp. 6, 132, 171):

> There must be some efficient cause for each slight individual difference, as well as for more strongly marked variations which occasionally arise; and if the unknown cause were to act persistently, it is almost certain that all the individuals of the species would be similarly modified. (Darwin 1875, p. 171)

The same applies to phenomena of extinction when conceived as part of the course of phenomena subjected to an order. And, as belonging to the intrinsic order of Nature, the "causal relationship" exhibits the universality reflected in some Darwinian epistemological requirements.

The recognition of a causal relationship implies the admission of the uniformity of causes for the uniformity of effects (Darwin 1875, p. 6). If similar organizations suffer an "action in a similar way", certain strongly marked variations frequently recur; if the existing conditions remain the same, the "mutant" will transmit to the descendants a tendency to vary in the same way; and if the "tendency to vary in the same way" is often so strong, all of the individuals of the same species will be similarly modified, without the help of any form of selection (Darwin 1875, p. 72). Several examples in the *Origin* show that a tendency to vary and be modified in the same way occurs when similar organizations suffer an "action in a similar way" (Darwin 1875, pp. 114-115, 125, 131, 132). Even when the causes are unknown, the similarity of effects leads to the consideration of a similarity of causes. The inference of similar causes from similar effects is guaranteed by the "uniformitarist" view (Darwin 1875, p. 125).

We may, however, be ignorant of the causes or conditions that determined this or that occurrence in particular situations. Or we may not be able to attribute the due weight to the causal factors involved. But these limitations do not impede the formulation of an explanatory pattern in terms of general conditions for the occurrence of a certain type of phenomenon. This pattern allows us to see what could be called the "causal structure" of that occurrence, although it does not enable us to point out the factor responsible for the causal efficacy in that particular case. For instance, we are ignorant of the causes that determined the presence of a few great quadrupeds in South America in spite of the existence of favorable physical conditions for that, and their abundance in South Africa. In spite of that, we do see that certain districts and times are much more favorable than others for the development of a quadruped as big as the giraffe, whichever might have been the causes (Darwin 1875, p. 179).

2.3. The causal structure

Several passages clearly show that it is a net of factors or group of conditions that are at play when we look for "causes" and the criteria for recognizing them:

> [...] if all the flowers and leaves on the same plant had been subjected to the same external and internal condition, as are the flowers and leaves in certain positions, all would have been modified in the same manner. (Darwin 1875, p. 174).

That net constitutes a "causal structure", within which the "effective causal factor" can be tracked, to use phrases of Sober (1984). This pattern of causal explanation applies not only to organic phenomena and, and in particular to those in relation to which "natural selection" comes on the scene, but also to phenomena such as the "glacial condition of the climate" (Darwin 1875, p. 336) and the "accumulation of the great fossil deposits" (Darwin 1875, p. 427). Concerning the organic phenomena, there are two fundamental factors that demarcate the structure in which "natural selection" is the predominant causal agent:

> In all cases there are two factors, the nature of the organism, which is much the most important of the two, and the nature of the conditions. (Darwin 1875, p. 106)

The structure on which the form of each organic being depends on comprehends countless and complex relationships: the appearance of variations due to causes too intricate to be followed, the nature of the preserved or selected variations, which, in turn, depends on physical and organic conditions, and finally inheritance (Darwin 1875, pp. 100-101). Also, the explanation of the very absence of evidence for the theory of natural selection depends on a group of conditions:

> These causes, taken conjointly, will to a large extent explain why – though we do not find many links – we do not find endless varieties, connecting all extinct and existing forms by the finest graduated steps. (Darwin 1875, p. 313)

This group itself is constituted by a varied cast that can include "circumstances thoroughly different" from each other (Darwin 1875, p. 263), "peculiar" (Darwin 1875, p. 113), "favorable" or "unfavorable" to the production of a certain result, and

> A grain in the balance may determine which individuals shall live and which shall die. (Darwin 1875, p. 411)

2.4. Causes as necessary and / or sufficient "conditions"

In the midst of so many factors that integrate the causal structure, recognizing the causal efficacy factor amounts to identifying the one to which a given occurrence or state-of-affairs is attributed or due.

Frequently, Darwin uses "cause", "effect" or phrases that designate the meanings of "cause" in terms of "conditional" statements (of the kind "if…, then…", "as…, then…", "given that…, then…"), or of a syllogism, whose premises contain the foundation for what is stated ("…hence…", "… therefore…", "… follows from…"). When Darwin does not explicitly make use of these statements or arguments of a syllogistic form, the passages where he uses those terms and phrases may be re-written in such a way that the antecedent provides the foundation for the consequent. In a broad sense, such an antecedent condition is that which "causes", "produces", "gives origin", "modifies", or "justifies" what follows it as a consequent.[6]

If we bring Darwin's phrases to a more standard approach to causal relationships, the performance of the varied cast of Darwinian "causes" can be seen in terms of "necessary" or "sufficient" conditions. Although Darwin himself uses the latter phrases several times, they bear particular meanings in Darwin's "one long argument". To begin with, a cause is always necessary. In certain situations some causes can thus be clearly identified. For instance, a variation comes as a condition clearly "necessary" but not "sufficient" for the action of "natural selection" (Darwin 1875, p. 356). In order to be "sufficient", it must be "useful" or "injurious". The exposition to new life conditions for several generations is "necessary" to produce any great amount of variation, but it is not sufficient, because it still depends upon the nature of the organism and on those conditions affecting the reproductive system (Darwin 1875, pp. 5-6). The identification of "necessary conditions" can also be made by inferring from a present situation, a former condition that would be "necessary" for the present occurrence. That is the case when we infer from the present conditions of the world that a past drop of temperature occurring simultaneously in the world during the glacial period must have happened (Darwin 1875, p. 336). From the present distribution of certain species we can infer their expansion from one island to anoother (Darwin 1875, p. 356).

---

[6] In Darwin, "causes" and "reasons" are interchangeable, which can be seen from the way he identifies "causes" and "reasons".

However, the usual evaluation in terms of "necessary" or "sufficient" is not helpful in some important cases. For instance: when examining the possible effects of the physical conditions on variations, we see that under dissimilar physical conditions similar variations occur, and that under similar conditions dissimilar variations occur. That means that different "causes" would produce the same effects, and that different effects would be produced by the same causes. Darwin refers to the nature of certain conditions as being of "subordinate" importance:

> [...] we clearly see that the nature of the conditions is of subordinate importance in comparison with the nature of the organism in determining each particular form of variation; perhaps of not more importance than the nature of the spark, by which a mass of combustible matter is ignited, has in determining the nature of the flames. (Darwin 1875, p. 8).

What is this "subordinate" importance? Certainly it has to do with a hierarchical interaction of factors. A "causal structure" is thus delineated, but one that escapes the traditional determination in terms of a group of conditions separately necessary and jointly sufficient for the production of the effect. The indication of what could be considered "sufficient condition" in Darwin's text is related to the guiding view of a "struggle for existence" on different levels (Darwin 1875, p. 411), as well as to particular factors involved in the production of new forms. By means of correlated variation the modification of a given part may be "necessarily" followed by the modification of other parts (Darwin 1875, p. 415). Concerning the view of a "struggle for existence", Darwin says that "natural selection" brings about extinction and divergence of characters (Darwin 1875, p. 103).

In spite of examples that would seem to be fairly adjusted to covering "laws" explanatory models, the larger number of cases in the *Origin* are those in which "sufficient" conditions also figure as "necessary" ones in the light of the contextual integrity found in the *Origin*. In light of this context, "conditions" are necessarily linked in a non-atomist (non-analyzable in a sum of separable parts) net of significances, whose threads weave them as sufficient and necessary. The need for such a net is imposed by the very definition of the central principle of the long Darwinian argument:

> I have called this principle, by which each slight variation, if useful, is preserved, by the term Natural Selection, in order to mark its relation to man's power of selection. (Darwin 1875, p. 49)

> This preservation of favourable individual differences and variations, and the destruction of those which are injurious, I have called Natural Selection, or the Survival of the Fittest. (Darwin 1875, p. 63)

Or, as already cited:

> Nature, if I may be allowed to personify the natural preservation or survival of the fittest, cares nothing for appearances, except in so far as they are useful to any being. She can act on every internal organ, on every shade of constitutional difference, on the whole machinery of life. Man selects only for his own good: Nature only for that of the being which she tends. (Darwin 1875, p. 65)

The meaning of "favorable" or "unfavorable" variation requires the consideration of an entire context of factors, which is represented in the view of nature as a "struggle for existence". Among other factors, there is the existence of places to be better occupied in the economy and politics of Nature, and the organic fact of inheritance. Without taking all these factors into consideration, there is not the required "sufficiency" to make "natural selection" to be the *vera causa*. As a *vera causa*, "natural selection" encompasses "necessity". In some situations, Darwin offers what would clearly seem to be a necessary and sufficient condition, for instance when he says that the divergence of characters "only" depends on the descendants of a species being capable of taking many and different places in the economy of Nature (Darwin 1875, pp. 303-304). Some other times, in which double condition is accomplished when a "necessary" condition is tracked from the examination of the effect to be explained, and, once this condition is reached, it proves to be "sufficient" for producing that effect.

Such relationships, in turn, require a semantic dimension that gives significance to the facts and their relational structure. This dimension is provided by the theory of natural selection as a whole. In several passages, Darwin asks for his theory to be evaluated on the basis of its explanatory power as a whole. In the conclusive paragraph of the *Origin*, the connection between Darwin's explanatory principles exhibits the character of "necessity" and "sufficiency", and integrates them around the explanatory power of the key image of the "struggle for existence":

> These laws, taken in the largest sense, being Growth with Reproduction; Inheritance which is almost implied by reproduction; Variability from the indirect and direct action of the conditions of life and from use and disuse: a Ratio of Increase so high as to lead to a Struggle for Life, and as a consequence to Natural Selection, entailing Divergence of Character and the Extinction of less-improved forms. Thus, from the war of nature, from famine and death, the most exalted object which we are capable of conceiving, namely, the production of the higher animals, directly follows. (Darwin 1875, p. 429).

We certainly find in Darwin's text passages where it seems appropriate to speak of causal relationship in the "traditional" Humean terms, as if "cause" and "effect" were two kinds of different and independent "phenomena" in such a way that whenever one of them comes to mind, the other one follows. That seems to be the case in the relationship between geographical and climatic changes as phenomena to be followed by changes in the geographical distribution of species (Darwin 1875, p. 321). Or, to give examples of another phenomenal order, the cases of changes in habits followed by inheritable effects (Darwin 1875, p. 8), or new life conditions followed by change of habits (Darwin 1875, p. 165). But even when apparently fitting the Humean pattern, the Darwinian view leads us to challenge the traditional Humean condition according to which "cause" and "effect" should be phenomena describable independent from one another. The production of "new" forms as forms modified and improved in relation to their ancestors is the phenomenon to be explained *par excellence*. The established causal relationship is translated in terms of the intrinsic and hidden bond uniting the "old" and "new" forms. The description of both terms, the "antecedent" and the "consequent", requires a semantic bond of such a nature that, if any descendant of a certain group becomes modified to the extent of losing all the traces of his/her parentage, he/she loses his/her place in the natural system (Darwin 1875, p. 370). The semantic bond between antecedent and consequent remains a necessary condition for the possibility of their relationship.

2.5. The "process-and-result" relationship as a causal condition

In the contextual integrity of the "one long argument" of the *Origin*, that bond finds a privileged expression as a "process-and-result" kind of relationship. "Natural selection" is a process in itself and the main cause (the condition *par excellence*) for the production of new organic forms in Nature. Whatever the concurrence of causes or conditions may be, they must be somehow conjugated and submitted to the action of "natural selection". Whatever the concurrence of causes or conditions may be, it is the solid accumulation of beneficial differences that gives rise to the most important modifications of structures in relation to the habits of each species (Darwin 1875, p. 132). Any amount of modification, says Darwin, can be generated by the accumulation of light, spontaneous, and useful variations (Darwin 1875, p. 233). This process of accumulation improves and converts varieties into good and distinct species. In this process "natural selection" equally tends to exterminate the parental forms, and the intermediate links or those forms that are more slowly modified and improved (Darwin 1875, pp. 134, 136, 138).

The relationship between "natural selection" as a process and the "survival of the fittest" as a result is inevitable, since that process necessarily leads to this result, which, in turn, as a result restricted to the scope of natural causes, can be produced only by "natural selection". Such a relationship does not follow from a mere correlation between a group of conditions called "natural selection" and a given result called "survival of the fittest". The process of natural selection is the process of production of fitter forms. The very concept of "fitness" is a relational one, a concept that depends on the same conditions and circumstances (delineated in the view of Nature as "struggle for existence") that provide the bases for the process of "natural selection". There is, therefore, a constitutional link that integrates the phenomenal reality in a net of significances and by virtue of which the phenomena gain their own identity. There is no way to think about the "process of natural *selection*" and to determine it as a phenomenal course without taking the result to be achieved, the "survival of the fittest" (although we cannot know in advance which particular form will be the fittest one), as its guiding thread.

The necessary link between "process" and "result" does not also imply that the Principle of Natural Selection is deprived of empirical significance, or that this principle exhibits a vicious circularity.[7] In its own formulation and for its intrinsically relational condition the Principle "regulates" (and so organizes) the conditions of experience. At the same time, the empirical conditions receive their significance from that Principle and support to its explanatory power "showing" its empirical significance. Such a bond between process and result does not require the existence of a plan to reach an "end" previous to and independent of the process. Both process-and-results cannot be thought of without one leading the other, and they are mutually conditioned by the same net of significances which makes the "real" and does not work in a linear way. The understanding of the way it works demands some consideration about the modes of causation to be found in the *Origin*.

3. Modes of causation: the confluence of the epistemological and ontological dimensions

The philosophical context Darwin dealt with[8] was concerned with "mechanical" causality, but also with causality in terms of (chemical) "transformation", "historical" and "final" causality, and with "causal chains". The first one is concerned with such a relationship that "causes" (C) and "effects" (E) are types of phenomena and that whenever C happens, E "automatically" follows. The mere presence of C in the course of phenomena determines the presence of E, without supposing any "internal" (and not "visible") bond between them. Their connections are merely "external" and related to "observable" and "positional" properties in the course of phenomena. There are several passages in the *Origin* that illustrate the pattern of a "mechani-

---

[7] A detailed analysis of the criticisms made to the Darwinian Principle in terms of "tautology" can be found in Elliot Sober (1984).
[8] This context is marked by the philosophy of John Herschel, William Whewell and Stuart Mill.

cal causality".⁹ In general, they refer to: the action of physical conditions, such as climate or other control factors that determine the average number of individuals of a species (Darwin 1875, pp. 54, 58); geological movements that determine the nature of the formation of geological layers, and the success or not of the fossilization processes; or climatic conditions, geological movements and displacement of icebergs that determine the migration, isolation, and communication conditions among the organic beings. Conditions "mechanically" produced supply the bases for the subsequent action of natural selection, as in the case of mimicry, when an individual's accidental similarity with another, protected group provides the basis for the acquisition of the most perfect similarity by means of the action of "natural selection" (Darwin 1875, p. 378).

Nevertheless, for the main concern of the *Origin*, namely, the production of new organic forms, merely "mechanic" causes are not sufficient:

> In the earlier editions of this work I underrated, as it now seems probable, the frequency and importance of modifications due to spontaneous variability. But it is impossible to attribute to this cause the innumerable structures which are so well adapted to the habits of life of each species (Darwin 1875, p. 171).

It does not "mechanically" follow from the fact that a part slightly varies that the necessary parts vary in the "right direction and degree" (Darwin 1875, p. 179). The "mechanic" causality has then to be over-determined by a "non-mechanic" causality, be it due to the fact that "natural selection" operates on mechanically supplied bases, due to the "ever-increasing circles of complexity" of the integration of several factors involved in the production (or elimination) of organic forms in their mutual dependence (Darwin 1875, p. 57).

What might this "non-mechanic" causality be? The strongest candidate, be it for the classic opposition "mechanic-teleological", be it for the indications found in the Darwinian text, seems to be a "teleological causality".¹⁰ In addition, a "teleological causality" is appropri-

---

⁹ For instance: Darwin (1875, pp. 58, 67, 68, 73, 82, 100-101, 111, 194, 198, 283, 291, 295-296, 299, 300, 319, 340, 350, 377, 383, 386, 394, 410, 414, 415, 420).
¹⁰ A "teleological causality" operates on the basis of the concept of an "end".

ate to the view of a *totality* in such a way that properties, processes, and state-of-affairs occur, and are determined by virtue of their "contribution" (many times referred to as "function") for the existence of the whole, or for the accomplishment of the preferential state / condition to which the whole as such is addressed (many times referred to as its "goal direction"). This "contribution" may be taken in relation to particular ends to be reached by the whole, as well as to the achievement of a "final end" which imposes itself as the regulatory principle of this whole or systemic totality.

Our ignorance on such a complexity is deep (Darwin 1875, pp. 100, 157). Our "traditional" and "simplified" access to the causality involved there is not enough to penetrate into the complex subject of the "origin of the species", and to understand the "the economy on any organic being as a whole". In the *Origin*, we find both modes of causation, "mechanic" and "teleological", and both are many times combined within a "teleological" framework.[11] This framework may internally hold phenomenal determinations according to a "mechanical causality". For instance, a chain of causes operating mechanically and teleological may be seen in the passage where Darwin argues that the geometric rate of increase of all organic beings leads to the full occupation of the territory, and the number of the favored forms is thus increased and the less favored ones become rare. Adding to that increase the occurrence of seasonal fluctuations, or the increase of the number of enemies, there is a good chance of the extinction of less favored forms (Darwin 1875, p. 85). The reference to "favored" and "not favored" forms, as well as the question about the occupation of places in Nature and the implicit "competition" that takes place there bring into light the background of a complex view of Nature as "struggle for existence", which is not reducible to a linear sequence of phenomena (Darwin 1875, pp. 230, 279). Variations of the right kind and degree (whose causes are for the most part obscure or unknown) must be given before "natural selection" can act on them, and grada-

---

[11] On the basis of a "mechanic causality" as well as in the case of a "teleological causality", there is a certain way of *conceiving* the nature of the "entities" involved in the cause-effect relationship. The way of conceiving them will differ in the one or in the other case, but in both cases we are dealing with *conceptions* about the way things are.

tions of structure beneficial to a species at each step of the formation process of species will only happen under particular circumstances (Darwin 1875, pp. 270, 180):

> [...] all spontaneous variations in the right direction will thus be preserved; as will those individuals which inherit in the highest degree the effects of the increased and beneficial use of any part. (Darwin 1875, p. 188)

3.1. Teleological causality

How could a teleological causality not be in conflict with Darwin's own words? When answering an objection that rose against his supposed attribution of "intentionality" to Nature, Darwin said he was just "metaphorically" speaking for "the sake of brevity" (Darwin 1875, p. 63). Also, more than once he referred to a "non-intentional" modification process or a process of "unconscious" selection by man (Darwin 1875, pp. 9, 80), that somehow builds the bridge to connect selection by man and "natural selection" (Darwin 1875, pp. 9, 25). Nevertheless, when talking about "natural selection" throughout his book, Darwin does not free himself from that metaphorical way of speaking. The elaboration and defense of the explanatory power of "natural selection" as a *vera causa* depend on such a way of speaking. It is a difficult the position Darwin is in, and it demands, among other things, a breakthrough in the view of what "intentionality" means.

Darwin seems to be aware of that difficulty when describing the construction of the beehives for the Melipona bee. Examining the formation of the architectural instinct of the bees for "natural selection", Darwin attributes to the bees the ability to judge precise distances and perceiving when they gnaw the wax until it has the appropriate thinness, and then they stop their work (Darwin 1875, pp. 222-223). But in the same passage Darwin explains the "automatization" of that judgment and perception. The "natural selection" would have come into the picture:

> [...] natural selection having taken advantage of numerous, successive, slight modifications of simpler instincts; natural selection having, by slow degrees, more and more perfectly led the bees to sweep equal spheres at a given distance from each other in a double layer, and to build up and excavate the wax along the planes of intersec-

> tion; the bees, of course, no more knowing that they swept their spheres at one particular distance from each other, than they know what are the several angles of the hexagonal prisms and of the basal rhombic plates; the motive power of the process of natural selection having been the construction of cells of due strength and of the proper size and shape for the larvae, this being effected with the greatest possible economy of labour and wax; that individual swarm which thus made the best cells with least labour, and least waste of honey in the secretion of wax, having succeeded best, and having transmitted their newly-acquired economical instincts to new swarms, which in their turn will have had the best chance of succeeding in the struggle for existence. (Darwin 1875, p. 227)

In the back stage, the borders between "intentional" and "non-intentional" are blurred, and a new view on the matter is slightly insinuated, as in the following passage:

> An action, which we ourselves require experience to enable us to perform, when performed by an animal, more especially by a very young one, without experience, and when performed by many individuals in the same way, without their knowing for what purpose it is performed, is usually said to be instinctive. But I could show that none of these characters are universal. A little dose of judgment or reason, as Pierre Huber expresses it, often comes into play, even with animals low in the scale of nature. (Darwin 1875, p. 205)

A small dose of judgment or reason is present even when the purpose of the action is not known. If the traditional distinctions among instincts, habits (non intentional sphere) and reason (intentional sphere) become more flexible and ask for a new approach to the issue, why not "go a step further" and question the usual commitment of a teleological view of the idea of a consciousness, a mind or an understanding where the ends to be achieved ("made real") are previously represented? Why could not intrinsic dispositions be put in the hardware? If so, the "metaphorical" way Darwin speaks of "natural selection" would not be a mere resource to style, nor would it commit any kind of violation of proper limits. This way of talking would just indicate the need to rethink some conceptions and distinctions.[12]

---

[12] I examine the innovative and cognitive role of metaphor in Darwin's thought in Regner (1997).

Considering the tone of Darwin's defense against the objections raised to his way of speaking of "natural selection", and the connotations that his use of this phrase confers to "natural selection" throughout his book, it seems that he is rather trying to protect himself from being blamed for advocating a certain sort of final causality or principle of finality (Darwin 1875, p. 63). Already in his Historical Sketch, Darwin criticizes M. Naudin's principle of finality because it would not explain the operation of Nature and would lead us to pseudo-explanations by requesting the interference of "supernatural" factors. According to Naudin, such a principle would be a mysterious and uncertain power – "fatality", for some, "providential will", for others – to determine, at all times, the form, size, and lifespan of each living being due to one's destiny, within the order of things to which one belongs. This determination would harmonize each being with the whole by virtue of the function or *raison d'être* one has to fulfill within the general organism of Nature (Darwin 1875, p. xix). Given the tone of Naudin's Principle of Finality, Darwin's criticism can be understood as a critique of a principle conceived in terms of "fatality" or "divine will". Such a conception blocks rational investigation and reduces the question about the production of new organic forms to a "predetermined plan" to establish which particular forms have populated, populate, and will populate the world.

In fact, according to the conception of teleology prevalent at Darwin's time, final causality seemed to be committed to the idea of a consciousness where the effect or end would be previously represented. Philosophically speaking, such an idea is tributary to the Cartesian "*res cogitans*" (intentional) / "*res extensa*" (mechanical) dualism, and to the Kantian *Critique of Pure Reason* dualism in terms of "matter / form" of our representations, which excluded the concept of "end" from the categories of "understanding" and thus from the objective knowledge of Nature, and restricted that concept to the domain of the pure concepts of "reason" while busy with itself. These philosophical views gave support to the teleological idea of a "plan of creation". However, these historical "commitments" do not exhaust the possibilities of teleological ways of conceiving Nature. Mechanical and teleological views of Nature are both just ways of conceiving it. In both cases, our access to Nature is mediated by "conceptions" whose cognitive value will depend on their heuristic capacity to satisfy their explanatory patterns, and make research advance. None of them can be said *a priori* as being more "objective" than the other. It is possible to conceive Nature as a system by attributing to it an intrinsic tendency to preserve its own identity as such a system. In this case, the principle of finality would be intrinsic to the hardware. If the Principle of Natural Selection is the central principle by means of which the system of Nature is self-preserved, that is, tends to its own end, that principle is internal to the system. There is no need to appeal to "providential will" or "fatality" to explain a principle of finality. Because of the "relational" nature of that principle within a system that makes room for the contingent and the complex "in ever-increasing circles", the idea of a successive plan previously determined for achieving particular ends is ruled out. In a special way, viewing Nature as such a system regulated by such a principle opens the way to rational investigation with its own "naturalism" without stopping at a "mysterious power", but accepting the challenge of showing *how* species are originated in Nature by pointing to a *vera causa*.

Besides the teleological nature we found in the very formulation of the Principle of Natural Selection, four points can be raised to guide the examination of a teleological causality to be found in the *Origin*: (a) the production / modification / acquisition of physical structures and mental qualities by the "accumulation of the variations in a given direction" appears in Darwin's comments just after offering definitions for the Principle of Natural Selection and in several other passages; these comments lead us to consider, on the one hand, (b) the idea of an "orientation" guiding the process" and, on the other, (c) the idea of "changes as adaptations", (d) in light of the idea of "totality" and the whole-part relationship in the structure of Nature and in the structure of the "long argument" that investigates it.

When attempting to elucidate the nature and causal force of "natural selection", Darwin frequently emphasizes the preservation and accumulation of favorable variations. "Natural selection" acts and accumulates useful variations as man does also (Darwin 1875, p. 34), but in the case of "natural selection" the criterion of usefulness is applied to what is useful to the bearer of the variations in light of his/her life conditions (Darwin 1875, pp. 65-66) pictured as a "struggle for existence". In this accumulative process, the more complex organs and instincts were "improved" (Darwin 1875, p. 404) and new and "improved" forms were produced (Darwin 1875, p. 84). This accumulation of light and useful variations affects, by means of correlation, other parts and results in a wide work of modification (Darwin 1875, p. 114) due to the continuous selection of individuals varying in the required way and degree (Darwin 1875, pp. 121-122).

The causal process of "accumulation in a given direction" does not follow from a mere sum of factors like the tendency of organisms to increase geometrically leading to a "struggle for survival", the occurrence of variations, preservation and heredity of useful variations, the tendency of variability to continue once it has started, and the occurrence of places to be better occupied in the politics of Nature. The Darwinian view of the "struggle for existence" (Darwin 1875, p. 50) as a picture of what goes on in Nature is not merely an item that joins the above ones, nor is it a mere result from that tendency to geometric progression. The net formed by several ingredients of that picture – such as the tendency of variability to continue and be "dominated by natural selection", and the decisive factor of the occupation of places in the politics of Nature – presupposes the view of a system as a totality bestowing an internal constitutive principle in relation to which the process of "accumulation in a given direction" can be thought. This process proves to be teleological in a double dimension. There is an "end" to rule the process as the "direction" that guides it. And this direction, at one level, is the direction of the "useful" variations to their possessor. On another level, considering the context as a whole, the "survival of the fittest" regulates the preservation of the system of Nature itself through its changing empirical configurations.

Still, the accumulation in a given direction is teleological in a third sense: in the sense of "improvement", of leading to a greater perfection, to "progress", to advancement in organization (Darwin 1875, p. 97). The intrinsically relational character of that improvement guarantees the possibility that the final result will, "in some few cases, be a retrogression in organization" (Darwin 1875, p. 201). It is not accidental, nor is it a matter of mere *ex post facto* verification, that in light of the "theory of natural selection" the preservation and accumulation of useful variations lead to more improved forms, and that the "new" species are "improved" forms. The criterion for the "usefulness" on the basis of which variations are accumulated and "improved" forms are produced points to the accumulation of what "serves to" their possessors to turn them "best equipped" to the struggle or "best adapted" to life conditions (Darwin 1875, pp. 164, 295-296, 299). It is a central issue for the *Origin* to show *how* adaptations and co-adaptations were "improved" (Darwin 1875, p. 48). "Improvement", therefore, is an issue at the heart of the explanatory condition of the theory. Darwin literally uses the word "improved" in order to characterize the process of production of new forms: there is no territory where all native inhabitants are so perfectly adapted to each other and to the physical conditions under which they live, that none of them can still be better adapted or improved (Darwin 1875, pp. 64-65); local varieties will not expand before being considerably modified and improved (Darwin 1875, pp. 274, 409); if some forms are modified and improved, the others will have to be improved in a corresponding degree, otherwise they will be exterminated (Darwin 1875, pp. 83, 296); new and improved forms will inevitably supplant and exterminate the older, less improved and intermediate varieties (Darwin 1875, pp. 266, 413); unless the other forms also become correspondingly modified and improved in structure, they will decrease in number or be exterminated (Darwin 1875, p. 139); a few ones can be frequently preserved by adapting to some peculiar life line or inhabiting some distant and isolated area (Darwin 1875, p. 296). Darwin also uses "improved" for the explanation of several particular cases (Darwin 1875, pp. 141, 144, 147, 157, 164, 170, 192, 200, 279, 283, 309, 398), such as the successive "improvement" of the tweezers of certain crustaceans. (Darwin 1875, p. 193), or the graduated perfection of the instincts of some species of Molothorus (Darwin 1875, p. 214), or the long and graduate series of the architectural instinct of bees by means of steps that are "all tending to the perfect plan of construction" (Darwin 1875, p. 226).

The teleological character of the process of production of new organic forms is clearly shown in the view of the produced changes as "adaptations". Darwin introduces the key question of the *Origin* – how are new species originated in Nature? – asking *how* the wonderful "adaptations" and "co-adaptations" we see in Nature are possible.

Darwin's answer initially refers to the changes operated by man as adaptations to his ends (Darwin 1875, p. 22). In Nature, "adaptations" are referred to: modifications in the structure, instincts, and habits of the young and adult individual or of their larval states (Darwin 1875, pp. 154, 156, 169, 177, 180, 213, 292, 389, 390-396); seeds and eggs in the view of their transportation (Darwin 1875, p. 359); species, in general, in relation to their life conditions like climate, or to their relationships with other organic forms and with their habitat (Darwin 1875, pp. 285, 330, 329, 356); groups or particular species, as referred to the adaptation of a group of fresh water productions traveling through the sea and reaching other areas of fresh water, or to the flat fish in relation to their life habits, or to the Galapagos Islands' birds who were adapted to fly from an island to another or to live in their own island (Darwin 1875, pp. 186, 344, 356, 357); and "adaptations" in Nature can also refer to particular co-adaptations, such as those of the birds of Wood Islands and the Bermuda (Darwin 1875, pp. 348-349).

The consideration of adaptations leads to the issue of "ends" or "purposes" as examples of the general rule of the same end being reached through different means (Darwin 1875, pp. 153-156, 165, 375, 414). Merely analogical likeness is an example of this rule (Darwin 1875, p. 374). But the same means can also serve to reach different ends, as in the case of the front members that once served as legs to a remote progenitor, and through a long course of modification came to serve as hands to some descendants, as oars to some, and as wings to others (Darwin 1875, p. 393). In several explanations at different levels and from different approaches Darwin literally refers to "purpose". The most surprising modifications can be perfectly "adapted for the same purpose", or they can serve the "same special purpose" (Darwin 1875, p. 170). A highly important fact, according to Darwin, is that an organ originally built "for a purpose" can be converted into another organ, "for a completely different purpose" (Darwin 1875, p. 148). For instance, the adaptation to the purpose of cutting herbs can explain the format of the beak of the common goose (Darwin 1875, p. 185). Differences in the larval and adult states of the same individual can also be explained by their serving different purposes at different stages of life (Darwin 1875, pp. 387, 391). The occurrence of rudimentary organs can be explained in a similar way (Darwin 1875, pp. 398-400). The adaptation to life conditions is the great referential for the adequacy or not of causal explanations of modifications in terms of adaptations to certain purposes. That is the "end" (Darwin 1875, p. 308) to which the other "ends" tend, be this tendency seen at the level of the ultimate encompassing end (or the "goal direction") of the system, namely, the "survival of the fittest", be that tendency seen at the level of the particular formats that ultimately end exhibits within the particular contexts acted on by "natural selection".

"Adaptation", however, is not a mere "answer" from organisms to their environmental conditions. Darwin talks about the best adaptation of the soil to a species, as well as about species being adapted to their own places (Darwin 1875, p. 356). He talks about the climate being perfectly adequate for certain species (Darwin 1875, p. 339), and about the South areas becoming adapted to the inhabitants of the North after the glacial period (Darwin 1875, p. 330). These considerations lead us to look at the relationships of adaptations in Nature from the perspective of Nature as a totality. The "adaptation" results from a relationship between the nature of organisms and their life conditions. The "purpose or end to serve" puts a given part in relation to the whole. The relationships between flowers and insects determine the morphology of flowers (Darwin 1875, p. 175). The correlation of variations determines the modification of parts other than the one that is modified (Darwin 1875, p. 415). The complex relationships of mutual dependence in the complex elaboration of forms are regulated by laws and factors acting around us (Darwin 1875, p. 429). In all the cases, the organic relationships among organisms in the struggle for existence are the most important of all relationships (Darwin 1875, p. 319). These and other phenomena lead to a causal relationship that escapes a strict mechanical pattern and direct our attention to a whole-part relationship subjected to "sets of conditions" and "concurrence of causes" as part of "causal networks".

"Causal networks" prevail even when we are faced with typical cases of mechanical relationships among phenomena, such as when Darwin tries to show that the glacial condition of climate is the result of several physical causes brought into the picture by an increase of the eccentricity of the Earth's orbit; and the indirect influence of this eccentricity on the sea currents seems to be the most powerful cause of that glacial condition (Darwin 1875, p. 336). The "networks" can hold internal segments describable as "mechanically" caused within a systemic teleological frame, which can be detected in several examples (Darwin 1875, pp. 51-56). Many times such a complexity cannot be reduced to a mere "action-reaction" relationship of a linear nature, nor can it be analyzed in terms of a mechanical "retro-feedback" of a causal chain. There must be an approach to each new stage from the perspective of the whole as such in light of which the new stage achieved affects the entire causal chain. The relationship that takes place among several parts of this chain is not reducible to a mere "reciprocity". Nor is it a case of mere "correction" or "reinforcement" by retro-feeding. In any case, a view in terms of "retro-feeding" would demand a view of a "program" belonging to the whole as such.

According to Darwin, the relationships among all organic beings are infinitely complex and they are fitted to one another and to their physical conditions of life. Consequently, infinitely varied diversities of structure can be of use for each being under changing conditions of life (Darwin 1875, pp. 60-62). And as any change in the numeric proportions of any inhabitants seriously affects the proportions of the others (Darwin 1875, p. 64), the cause of the decrease of any species depends not only on its own injury, but also on another species being favored (Darwin 1875, p. 54). In his explanation of the facts related to Classification, Darwin reaffirms the occurrence of "almost endless circles" in the relationships of organic beings with each other and to their "conditions of life" (Darwin 1875, p. 104). Darwin does so at different moments of his "long argument" and at different explanatory levels, from the complexity and particularity of the explanations of organic phenomena when compared with explanations of physical phenomena, to the explanation of the development of delicate coral-lines (Darwin 1875, pp. 55, 57, 58, 73, 100-102, 135, 157, 304, 314, 387, 402). In his all-pervading reference to "co-adaptations" (Darwin 1875, p. 48), Darwin emphasizes that almost every part of any organic being is so "beautifully" related to its complex life conditions, that its sudden and perfect production seems as unlikely as the invention by man of a complex machine that would be already perfected (Darwin 1875, pp. 33-34). The "improvement" is a process.

3.2. A view of Nature once again

The view of a way-of-being of Nature as the basis for thinking Darwinian causality comes from the need to account for the relationships of the organic beings among themselves and in relation to their "conditions of life" from the perspective of a contextual integrity, that is, a "complex totality" represented by the picture of the "struggle for existence" (Darwin 1875, pp. 49-50). We are led to an ontological dimension when we look for the foundation of the epistemological dimension in order to examine the Darwinian causality. The complexity of that whole is not only due to its countless and mutually interfering constituent factors, but also to the nature of their relationships in "circles of involving complexity", escaping the usual patterns of analysis.

In light of that picture, its guiding unity principle, namely, "natural selection" or the "survival of the fittest", plays the role of an integrating causality that interacts with other factors, such as inheritance and the occupation of places in the politics of Nature (Darwin 1875, pp. 49, 270). Its integrative role implies the general frame to order the process in terms of an "end" that belongs to the system in its relational integrity, and so guarantees the preservation of the system through the re-accommodations and modifications of the different moments of its empirical configurations, without predetermining which particular empirical configuration the system will come to present. The inherent circumstantial character of the empirical determinations of the process is part of the system whose general teleological framework allows sheltering and recognizing segmentary and non-teleological relationships within the system. Not everything will be teleologically produced, although the Principle of Natural Selection "overdetermines" all relationships. A teleological view imposes itself as a frame to the general picture whose detailed determination, which involves "mechanical" and "teleological" causality, makes it possible to show the mechanism that is not a matter of blind necessity. There is a "direction" to the process that is intrinsic.

In its first definition, Darwin presents Nature as "the aggregate action and product of many natural laws", and we are led to treat Nature as a "system" of laws, but without autonomy and as an "object" to be ascertained by us (Darwin 1875, p. 63). Soon afterwards (Darwin 1875, p. 65), Darwin introduces a second definition of "Nature" by identifying it with the "natural preservation or survival of the fittest" and seeing Nature as acting on "the whole machinery of life", selecting, exercising the selected characters, etc. In other words, Nature is then presented to us as a "subject" endowed with "autonomy". Although Darwin does not speculate on the relationship between his two closes and apparently conflicting definitions of Nature, we may attempt to reconcile them keeping in mind Darwin's versions of his main question, namely, "how are species originated in nature?" How are adaptations and co-adaptations produced? How do varieties become good species? How do species form genera? How do groups become groups within groups? (Darwin 1875, pp. 48-49), and we may do it keeping in mind the complexity of the involved relationships as well. The attention given to these points allows us to understand the meaning of causality as contained in the Principle of Natural Selection in terms of an internal principle to Nature as an "autonomous" system (according to the second definition), but susceptible to empirical determination (according to the first definition). Starting from such a view of Nature, the *Origin* manages to reconcile two tasks whose accomplishment opens the way for thinking about teleology in new and contemporary terms.

References

Darwin, C. (1987), *Charles Darwin's Notebooks, 1836-1844* (edited by P.H. Barret *et al.*), Ithaca: Cornell University Press.

Darwin, C. (1896), *Life and Letters of Charles Darwin*, 2 vols., Darwin, F. (ed.), New York/London: D. Appleton and Co.

Darwin, C. (1875), *On the Origin of Species by Means Natural of Selection or the Preservation of Favored Races in the Struggle goes Life* (from the 6th English Edition, 1872), New York: Appleton.

Herschel, J. (1966), *A Preliminary Discourse on the Study Natural of Philosophy*, New York/London: Johnson Reprint Corporation.

Himmelfarb, G. (1968), *Darwin and the Darwinian Revolution*, New York/London: Norton & Co.

Mill, J.S. (1843), *A System of Logic*, 2 vols., London: John W. Parker.

Regner, A.C. (2001), "O conceito de natureza em *A origem das espécies*", *História, Ciências, Saúde–Manguinhos* VIII (3) (23): 689-712.

Regner, A.C. (1997), "O papel da metáfora no longo argumento da 'Origem das Espécies'", *VI Seminário Nacional de História da Ciência e tecnologia–Anais, 1997* 6: 35-40.

Regner, A.C. (2001), "Padrões explicativos darwinianos", in Suliane, A. (ed.), *Etnias e Carisma*, Porto Alegre: EDIPUCRS, 2001, pp. 74-90.

Sober, E. (1984), *The Nature of Selection*, Cambridge: The MIT Press.

Whewell, W. (1967), *The Philosophy of the Inductive Sciences Founded upon Their History*, 2 vols., New York: Johnson Reprint.

Darwinism in the Mid-20th Century and the Relations between Philosophy and Biology[*]

Eduardo A. Musacchio

National University of Patagonia San Juan Bosco (UNPSJB)

1. Objectives of the present contribution

In the Theory of the Evolution, such as it is presented in texts of recent Biology or Paleobiology, it is possible to recognize a series of transformations as a result of successive contributions. The retrospective observation is useful in the sorting process of historically stable components of the theory and the nature of the modifications. In order to clarify these changes, it seems pertinent to compare the series of transformations in Biological Evolution against those stated in the Theory of the Cortical Evolution, both proximate by historical reasons. In each of the series of transformation three periods are distinguished: foundational, unifier and instrumental (table 1). Several analogies between both theoretical networks allow a better discerning of

---

[*] The author thanks to Dr. Margarita Simeoni (UNPSJB-Argentina) the reading of the original, and to Dr. Romeo Cesar (UNPSJB-Argentina) the discussion of some topics of the present contribution. Dr. Anna Carolina Regner (UNISINOS-Brazil) made a valuable critic as referee to the original manuscript. However, the author is the only one in charge of the failures that the reader surely will find in the work.

the changes in scientific thinking throughout history, therefore reflected in each scientific field. These modifications are related to the different ways in which scientists proceed in order to research, generalize, corroborate, reject and explain.

The present paper is focused in the difficulties that the Theory of Evolution presents during the period that includes the Second World War and its consequences. During that specific period of time Darwinism is considered the most outstanding and accepted theory in the field of biological evolution. Authors who come from different fields of study share relevant aspects of the theory, characterizing an updated vision of the well-known biological change known as Neo-Darwinism (representative works: Mayr 1963, 1982, 1988, Simpson 1944, 1953, 1961a, Dobzhansky 1966, Ayala & Dobzhansky 1974).

It is during the mentioned period of apogee when some difficulties arise and disrupt Darwinism. North American paleontologists G. G. Simpson noticed at that moment different problems, four of which are listed below:

- The difficulties involved in choosing between if the change is a continuous or discontinuous phenomenon (in Simpson 1944; also see preface in the 1984 edition).
- The conflictive notion of species within the framework of fossil record (in Simpson 1961b).
- The directionality of the change phenomenon (in Simpson 1954 and in Simpson 1944, Chapter V).
- Finally, and related to point 3, the proposals of Rosa (1900) and Fechner (see Hennig 1968, pp. 298-302), on the progressive narrowing of evolution within the framework of the fossil record.

The author has believed to find in the pioneer biologists the presence of philosophical notions that come from their cultural education, which are reflected as departure points, accepted as primary terms. Among these are the notions of species and hierarchic system, both grounded on Aristotle's Categories and their ontological vision. In some cases, the biologists notice the presence of this philosophical tradition and fight to separate from its presumed consequences; such as the case of idea of teleology. This anchorage on "ancient" Greek philosophy seems to be the source of some difficulties in Darwinism and Neo-Darwinism (see below), which are analyzed in the present contribution. On the other hand, the elements of the stable nucleus link with the "modern" philosophical thought prevailing in sciences during the 18th Centuries; among these are mentioned: the population analysis, the natural selection and the significance of the fossil record.

2. Is it in conflict the notion of evolution in biology?

2.1. The permanent aspects

When doing an examination, which always turns out to be incomplete, of the wide biological literature for university education or for divulgation in the past decades, it's possible to affirm that the scientific community of biologists and paleobiologists acknowledges today, without reticence, the notion according to which the living beings come from sets of pre-existing individuals possessors of different attributes. The author does not know active biologists or paleontologists who reject the notion of evolution in Biology.

Among the different evolutionary designs that paleontologists recognize in fossil record, the transformation and the cladogenesis should be considered in detail. The design of *transformation* represents a succession of ancestral-descendants populations (or lineage), as the example illustrate in figures 1, taken from Bettenstaedt (1962 and 1968). In this case a single chrono-stratigraphic sequence of fossil-

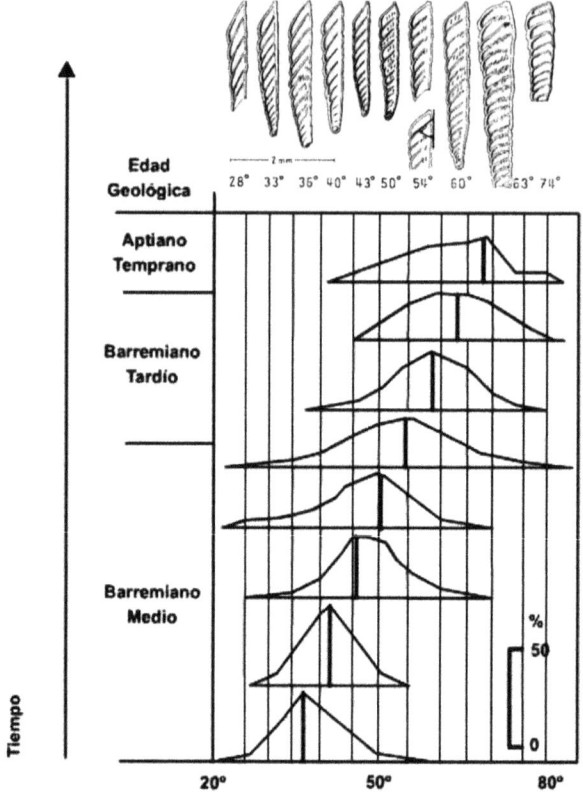

Figure 1. *Transformation* in a single lineages of Cretaceous foraminifers (microfossils of Protista) from Niedersaxen, Germany. During the process of change, the number of taxa remains. In *Vaginulina procera*, a succession of fossil "populations" formed by ancestor-descendants, Bettenstaedt (1962, Abb. 4-5) measured statistically the angle formed between septa and axis of the test. The average of the angular value increases continuously as the evolutionary tendency unrolls itself.

populations of Cretaceous foraminifers (Protista) exhibit a trend of changes in the slope of the chamber-septa related to the lengthening of the test. This tendency is statistically studied in different allocrhonic populations.

The second evolutionary design represents the branching evolution or *cladogenesis*, shown in fig. 2 based on foraminifers, also taken from Bettenstaedt (1962, fig. 13; 1968, fig. 8). In this case, a new lineage arises as a branch of the relatively stable axis of ancestral populations.

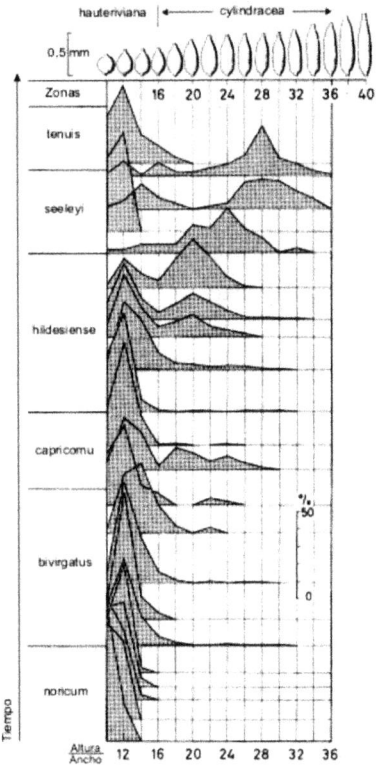

Figure 2. *Cladogenesis* in lineages of the Cretaceous foraminifer *Lagena hauteriviana* from Niedersaxen basin, Germany. The final result of the evolutionary process is duplication in the number of taxa. In this cases Bettenstaedt (1975) measured the relationships height / wide (L/T) of the test. The evolutionary tendency shows the appearance of a lateral branch of populations. The latter exhibit tests of progressive cylindrical form which separates from a stable lineage of "populations" with spheroid tests.

The two different evolutionary designs represented respectively by the figures 1 and 2, exhibit cases of continuous (or gradual) succession of populations. In both plains, the slow changes in the geological time are presumably due both to the low evolutionary-rate of the fossil group choice for example, as well as the continuity of the stratigraphic record. They are also well known numerous cases in the fossil record that show evolutionary designs similar to previous models of transformation and cladogenesis, however interrupted by gaps in the sequence of populations. These intermittences can be explained by different arguments. In cases, some external factors (environmental or stratigraphical, among others) "hide" the missing-interval in a lineage which evolved, however, with a regular rate of change. In other cases, some internal factors (chromosomal changes or regulatory-genes modifications, among others) produce sudden morphological modifications in the descendents breaking the continuity of the lineages. Such is the case represented by the succession of different microfossil taxa of *Cytherelloidea* (Ostracoda-Crustacea), associated with the above-mentioned foraminifers, in the same Niedersaxen Basin (cf. Bettensetaedt 1958, p. 117, Abb. 1). Differing of the pictures corresponding to the forams-lineages, which exhibit continuous changes, these ostracods show a punctuated or disjoined biostratigraphical distribution with relatively quick changes, this last one suggested by the taxonomic diversity of their carapaces.

For many naturalists, biological evolution is not only a coherent theory. In addition, the evolutionary phenomena in the nature can be observed, repeated or tested by an observer. In agreement with this affirmation, those naturalists could invoke Darwin (1859), Chapter 1: "The Variation in the Domestic State". The mentioned author observed distinctly the improvement in the descendants of animal populations and domestic plants that breeders obtained by means of selective crossbreeding of progenitors. Another argument could be the genetic melioration that biotechnologists practice currently to obtain vegetal varieties for economic profit. Both arguments above stated form part of a series of strategies that allow them to test the evolutionary phenomenon, which can be documented and explained by an observer, or a scientist, during the development of an experimental research project.

As Darwin and Wallace observed, natural selection is considered today an important factor of evolution and its reputation remains appraised, or without decrease in its recognition. Likewise, there is a consensus to assign to the biogeographical design a relevant role in the differentiation of populations that developed in isolation. The concept of fitness could be quantified in terms mainly of reproductive viability and amount of descendants in order to evaluate the effectiveness of different evolutionary factors in the descendants (see Sober 1993, p. 57). However, it must admit that the change phenomenon is more complex of the thought thing by Darwin and that it corresponds to treat it like a polynomial equation in which different evolutionary factors take part, according to the cases. Among these, the most important are: natural selection (already mentioned); periodical global crises of ecosystem; size of the founding population; genetic mutations; chromosome mutations involving the active role of genes in whole range; genetic drift, and the possible consequences (at least in theory) of early sexual maturation (paedomorfosis) on evolutionary forestalls appearing in juvenile individuals.

2.2. The nucleus of the theory of evolution and linked hypothesis

Together with the central notion of evolution of the living beings (nucleus: see terms 1-5 below) several hypothetical models are formulated, including among others the category species, the evolutionary design, the origin of the evolutionary units in nature and the evolutionary rate of the lineages. Many of these hypotheses have been long time debated and constitute the main source of data to construct the history of this theory. The central conception of biological evolution (nucleus) invokes the existence of: 1: Ancestors; 2: Descendants; 3: Relations of ancestors-descendants; 4: Ramified generation of sets of individuals, hierarchically subordinated to other ancestral ones; 5: Evolutionary factors that turn out to be effective in the transmission of new states and characters in the descendants, and to assure the preservation of the descendants. Term 4 in particular, implies that the generation must branch out; that is, be opened in the sense of time and not reticulated like in the tree of Porphyries (Sober 1993, figs. 6.1 and 6.2; see also Eco 1983, Chapter 2).

Regarding the second part of the theoretical corps (linked hypotheses and secondary theories); the main interest of this paper focuses on the category species, including different proposals on its nature, origin, transformation and evolutionary design. These divergences have generated several inaccuracies and conflicts that cloud the coherence between theoretical premises and empirical data shown by the grounds of the evolutionary theory. The disagreement between the various models on the concept of species has an epistemological substratum that will be later analyzed (see below in 4. The philosophical substratum). Right now it seems of interest to underline that a coherent theory dealing with the change of the organic world can be elaborate overlooking the concept of species. In this case, populations or individual organisms could be playing the required role of operative units in the evolutionary process. Darwin believed in the existence of species in nature. If the species were only nominal or conceptual units, how could they have originated themselves? And how can the new species being originated if the natural selection, controlling the diversity, is not enough to increase the latter one?

The present paper deals with the difficulties that arise in some explanatory self-defined Darwinian models formulated in the mid twentieth century, but focused from the point of view of the philosophy of science. Among these models, the outstanding and restrictive thesis held by Neo-Darwinism states that the source of variability between organisms is provided by the genetic mutations that happen randomly and the genetic recombination; therefore the phenomenon of controlled change being entirely generated by natural selection.

The permanence of ancient theoretical nucleus, foreseen by Lamarck among many others, and magnificently stated in the XIXth Century by Darwin; is not declined by the difficulties displayed in some additional parts or derivate terms that characterize the different models (including the one by Darwin/Wallace on the evolutionary species) or other evolutionary factors. In defense of the stability of the theoretical core despite the eventual falling of additional hypothesis, it can be argued better, comparing the case with that of cortical mobility in Earth sciences. The latter exhibits the permanence of a central nucleus (cortical tangential movements producing the paleogeographic changes) in spite of inconsistent hypothesis, which the theory itself has been discarding throughout time thanks to the advancement of knowledge.

It is also appropriate to consider the question that if the changes experienced by the Theory of Evolution constitute a true unfolding or extension of the original ideas exposed by Darwin. That is, if the principles exposed about natural selection controlling the appearance of new species give account of the evolutionary phenomenon that shows the fossil record. In the opposite case; if the set of primary terms must be reviewed, then we are facing a very different circumstance and the Theory itself must be reformulated.

In order to give an answer to this puzzle the most likely method appears to be a roundabout one. In the following section, the design of the theoretical network on cortical mobility in Earth Sciences will be dealt with, in order to benefit from the compared analysis between both disciplines.

2.3. Comparing theories of geologic change and fossil record

The debates on paleogeographic changes through geologic time and the involved tectonic movements that took place during first half of the twentieth century were useful in order to eliminate misconceptions. In addition, new contributions such as the study of oceanic bottoms at the beginning of the second half of the 20th Century, allowed them to consolidate the mobilistic conception.

The outline that follows compares the states of transformation in the two theories of Cortical mobility and Biological evolution respectively. The three states of historical development (a^0, a', a'') are only approximately contemporary between themselves (taken from Musacchio 2003, slightly modified):

	Cortical mobility	**Biological evolution**
19th Century (late) to 20th Century (early)	MC a^0 Continental drift	BE a^0 Early Darwinism
20th Century (mid)	MC a' Geosynclinals theory	BE a' Neo-Darwinism

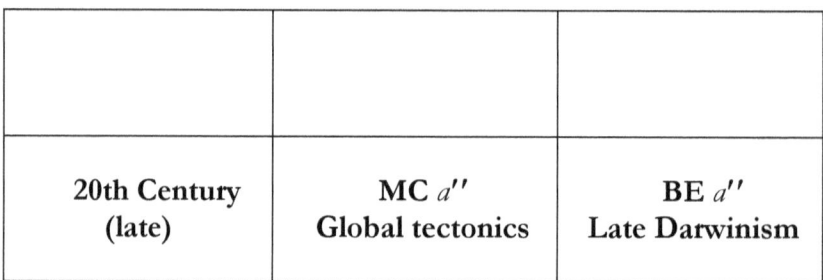

20th Century (late)	**MC** a'' **Global tectonics**	**BE** a'' **Late Darwinism**

Table I: Comparing stages of development between the theories.

It seems that geologic processes, which are recurrent and explainable in terms of physical-chemistry, differ from the change processes performed in organisms (Musacchio 2004). The author has also admitted that fossil record is characterized by missing forms of life, which denotes its historical nature. The latter feature cannot be deduced or explained by laws ruling the geological process. It is the same to say that regularities postulated in paleobiology are different from those in geology. This is not an obstacle for the recognition of several cases of necessary interdependences between both factual fields. Such are the cases where sequences of faunas or floras are in correspondence with the geological modifications of the physical environment. In addition, it must be taken into account that the bio- and litho-stratigraphic record is shared in both disciplines. To this last one condition is related the "measurement" of geologic time on the base of the irreversible changes of the fossil registry (Musacchio 2004).

For the purpose of the present contribution, it seems more pertinent to focus on the analogies (or isomorphisms) between both theories than to consider the factual interdependences; in particular those exhibiting correlative historical periods within the respective networks. These isomorphisms can be explained analyzing the notion of scientific theory that prevails in each field for its respective historical period. For the present case, it seems that changes in the scientific thinking are determinant factors when they compare with the significance of new research results in order to deal with the conceptual transformations of each discipline.

The next paragraphs display the periods of historical unfolding: $a°$, a', and a'', forming part of the Cortical Mobility network. They are followed by comparisons with the Theory of Evolution in Biology and the search for possible isomorphisms.

aº: Continental drift. At the beginning of 20th Century, Taylor (USA, 1908) and Wegener (Germany, 1910) recognized the occurrence of geographic changes in the geological past, resulting from the partition of a denominated Pangea Supercontinent (see Holmes 1952). However, an important previous notice about the paleobiogeographic design was outlined by Sinder (France, 1858; cf. Holmes 1952 and Holmes & Holmes 1980). The pioneering authors were impressed by the fit that could be obtained if some coastal margins were coupled, like those of Africa and South America from the appeared Atlantic edges in the maps. After Taylor, the effect of the lunar attraction when separating of the mother Earth and being located like its satellite – an event supposedly happened during the Cretaceous – was the cause of the displacement of the continents. After Wegener, however, the cause of the "pole fight" of the Supercontinent was the gravitational attraction exerted by the equatorial Earth swelling. This phenomenon was originated by the centripetal forces generated by the Earth rotation on its axis. The schematic argument of the continental drift theory, including some of the outstanding failures early issued, is squeezed as follows:

p (drift) implies q (gravitational attraction) (Taylor 1908, Wegener 1910)

    Auxiliary hypotheses:

Coupling of some continental margins.
Geological continuity in the lateral conformation of Africa and South America.
Relatively recent opening of the Atlantic
Free interchange (during the geological past) between Africa and America

    Non-experimental test of the Theory (Lambert 1921, Epstein 1921, in Howell 1962).

p (drift) ¬ (sufficient attraction)
-- (*"Modus tollens"*)
Not being the case of q it is not the case of p

This way, the hypotheses of Wegener are refuted some time after their presentation (Epstein 1921, Lambert 1921, in Howell 1962, pp. 280, 413 and 417).

The hypothesis of Taylor can be rejected in a similar way. During the three following decades, a great amount of geologists from the North hemisphere misestimated the contributions of Du Toit and other paleontologists and geologists working in the South hemisphere. The latter were comparing analogies originated by regional geology and the intercontinental biostratigraphy. In Europe and North America the belief prevailed on the structural permanence of the geographic design, beyond periodic sea-level swings in the layout of the coastal lines. At the time, the data of Paleontology are not being "confirmed" by the Geophysics, a discipline supported by the Physical Sciences.

a': Convective currents in the Earth's mantle theory and geosynclines. By the end of the 30's, Holmes is a representative author of ideas that will crystallize during the period of general acceptance of mantle convective currents Theory in the frame of Geosynclinals, towards the 60's. Holmes (1952), and other authors (see Holmes & Holmes 1980), will invoke the internal stresses operating in the Upper Earth Mantle as causal factors for the continental drift. The present historical stage is characterized by the search in Earth Sciences, of comprehensive explanations of real phenomena, their causes and interrelations. The notion of progress in knowledge (from: *progr☐ di* = to walk ahead) can adjust to the present situation. This vision can be, tentatively, formalized within the framework of the "truth" and "successive approach" discussed by Popper (1963), in which he expresses that the notion of progress, or approach to the truth of scientific knowledge, is based on a strategy of conjectures and refutations. Popper later incorporates these ideas to his more rigid theory, exposed in his *Logic of Scientific Discovery* (1934). The criterion of truth is also analyzed by Popper in the 60's inspired by the contributions of Polish author, Tarski. The geologists of the 60's seem to accept that their explanations, statements, and generalizations are in correspondence with the facts occurred in the geologic past. This notion, which is a realistic one, seems coherent with Popper's criticism to the notions of relativism and instrumentalism in his philosophy of science (Popper 1963). According to Popper, scientific theories should not be utilitarian instruments to find explanations or applications, but the result of a search for coherent knowledges. This last approach fits appropriately with the historical development of Geosynclines Theory, which adjusts comfortably to the general symbolic expression written down by Popper (1963, pp. 285-286):

$$VA\ (a) = Ctv\ (a)\ \square\ Ctf\ (a)$$

The verisimilitude, or degree of similarity of a theory to the truth [VA (a)], is in relation to its content of truth Ctv (a) reducing its content of falseness Ctf (a).

a": *Global tectonics.* An historical overview of the geological research during the second half of the 20th century allows it to recognize in the discipline two methodological attributes of approach. One is the development of multidisciplinary programs or research with shared objectives in order to clarify conflicting topics or shadowed areas of knowledge. Such is the case of the Deep Sea Drilling Project for the study of oceanic bottoms. Another one is the frequent use of models. As a result of the developed programs at the end of the 50's, evidences on the previous union between continents that formed Pangaea will be, finally, unquestionable. The scientific community of geologists, respectful of the approval that Theory obtains now from the field of the Geophysics, comes to admit that continents have fragmented and separated from each other. The Theory of the Global Tectonics (TG), or Plate Tectonics Theory, is becoming a multidisciplinary, encompassing and coherent "paradigm". Based on the principles of Earth Sciences of uniformitarianism and ciclicity, the theoretical nucleus of the TG can be outlined according to the following proposals:

1. The Earth is differentiated in vertical sense in layers of relative density and fragility (or ductility).
2. The upper part (or the lithosphere = crust + upper mantle) becomes geographically differentiated in plates – or blocks – relatively rigid.
3. The plates are exposed to forces originated in the interior of the Earth which determine its mobility.

4. Mobility is resolved in three types of tectonic activity: divergent, convergent and transcurrent (or in couple).

Within the framework of the TG, numerous geotectonic models have been proposed; sedimentary basins, bioprovincial differentiation and petrologic evolution, among others. These models are constructed from specific cases. However, they are permanently tested in order to process the data gathered experimentally (always incomplete), based on which the geologist tries to reconstruct the facts that took place in the past. The structure of the TG, such as Suppe (1998) displayed, is useful to illustrate an example of structure in the framework of the semantic conception of the theories (see Díez & Lorenzano 2002). The four proposals above designed with the symbols [H1-4] are considered by Suppe, although in a somewhat different order and content. A fifth term of the theory [H5] proposed by Suppe (1998) has some different epistemic attributes from the other four [H1-4] and was not included here.

The isomorphisms. Like analogies in Biology, isomorrphisms in theories show comparable or interchangeable formal structures, lacking of relations of kinship or dependency in the ontological level. As an example, analogies can be recognized between societies of insects and of human beings. However, they are not true homologies because both groups have not parental relationships; for this reason, in each ones of the mentioned groups, social attributes appeared independently. In our case, isomorphisms between correlative intervals of the theoretical networks (MC and BE) do not respond to connections between factual contents (or "substantial" components) shared by the disciplines. Merely, they seem to reflex the formal language adopted in every period by the respective scientific communities to discover, rule, confirm, refute or explain puzzles; illuminating in this way the scientific thought of the time. Let us characterize the three cases:

a° Foundational. The naturalists believe in the real and independent existence of the "facts" they describe. Affirmations that occur within the framework of the realistic-Aristotelian notion of theory in this period, normally respond to the *modus ponens* argumentative structure. Likewise, theories can be naively refuted (and it has in fact happened at that time). The Aristotelian structure of *modus tollens* set up the alluded isomorphism. The falsification of the theory also can be displayed in a propositional argumentative outline, as the squeeze below. Here follows an ingenuous refutation of the foundational Darwinist Theory, whose statement looks analogous to that one used above, in order to falsify the Theory of "Continental Drift": *"The natural selection proposed by Darwin is insufficient to explain the evolutionary novelties. Then, the Theory is wrong."*

p (new species) implies q (natural selection)

   Auxiliary hypotheses:

Succession of ancestor-descendents populations in the fossil record.
Changes produced by the breeders of domestic populations

   Genetic explains that:

p (natural selection) ¬ q (new species)

--("*Modus tollens*")
Natural selection does not generate new evolutionary features

That is to say: Not being natural selection a generator of evolutionary novelties, is not the case of the biological evolution.

a": *Comprehensive.* After the criticism made to foundational Darwinism by geneticists, the first stage is pursued by a period of all-embracing approach that undertakes the explanation the *quid* of the evolution. The improvements seem to be the result of conjectures and refutations. The theories continue to be realistic. In diverse factual disciplines belonging the present sciences, the limitations of the bivalent logic are noticed. On one hand, the statistical nature of the regularities contributes to the necessity of a structure of reasoning different from that one in propositional logic. In addition, the admission of the polynomial characterization of the processes stresses the necessity of a logic that also allows deviations at the Principle of the third excluded. In other words: a logic that admits different "degree of truth". A renewed interest by the Bayesians arises and different systems of multivalent logic are proposed. An interesting mark for this period is the recognition of the presence of the platonic thought (typology) in the pre-evolutionists biologists (see Mayr 1963).

a": *Instrumental.* Theories make possible a coherent interpretation of the field data. Strategies of analyses based on the Cantor's sets, the notion of supervenience and the trace of necessary interdependences (homologies) offer conceptual instruments for the comparative study between theories and their linkages. These strategies extend the exhausted possibilities of the opposed pairs reduccionism / emergency. In the respective disciplines, the formal representation of field data are characterized by the fruitful construction of models, which are useful as instruments for testing and comparisons. The stratigrapher, studying a local sedimentary basin, has conceptual models based on chosen cases, useful to explain and predict. On the other hand, the biostratigrapher analyzing a succession of ancestor-descendent populations has different models of "speciation" applied to provide meaning to the biostratigraphic record. Sometimes, it is possible to interchange models between the two disciplines. The hypothetical case proposed by Great (1989) can be shown as an example. In the later, dendrograms corresponding to the successive fragmentation of a super-continent can be shared with the one that designs the evolutionary changes of biota that inhabits successive fragments. The two variants of the same design allow invoking connectability and certain predictability in the different processes, taking into account the necessary dependence of the biota to the geography. However, the condition of interdependence shown by the previous example is not demanded to validate an isomorphism.

3. Some difficulties in Darwinism (evident conflicts and proposed solutions during the mid 20th century)

Early in the work *Tempo and Mode in Evolution*, Simpson (1944) had exposed several conflicts that the fossil record brings to the Darwinist vision of change. This work, today relevant within the field of paleobiology, was conceived to give "*ad-hoc* explanations" for difficulties, finely distinguished by the American paleontologist. However, the reading of this book transmits to the reader the conflicts that the mentioned author just tries to clear-up. This section deals with some of the subjects discussed in *Time and Mode* but arranged in the following order:

- Continuous processes versus discontinuous processes.

- The notion of species within the framework of fossil record.
- The sense of change and the progressive reduction of variability in the evolution.
- The notion of increasing complexity.

3.1. Continuous processes versus discontinuous processes

The opposition between *continuous processes vs. punctuated* (saltational) processes admit an ample range of historical visions disputing each other. Some of them express in a more or less intolerant way respect to that representing a supposedly opposite position. It seems adequate to consider altogether the aspects dealing with evolutionary rate, the main topics in the discrepancies. The controversies affect the geologic evolution of the terrestrial crust as the biological evolution (some considerations on this subject in Musacchio 2001, section 3.1). The Theory of Catastrophism (Cuvier 1817) has deserved in the two last decades a renewed interest. Undressed now from the Creationism shared by the pioneers, the renewed version admits that the "catastrophic" processes, causes of the evolution by jumps within the terrestrial crust, are not supernatural (see Ager 1993). On the other hand, Uniformitarianism (Lyell 1830-1833) ~ Actualism, (Hutton), tries to explain the geo-historical phenomena without taking into account supernatural causal factors. It seems important to emphasize that Lyell did not ignore the importance of the impact produced by paroxysms like volcanism (Lyell 1830-1833, p. 85ss.).

Despite the previous clarifications today it seems fragile to maintain a uniform rate for the geologic change. This one is a well-founded criticism to the hypothesis of unrestricted gradualism. Currently, we register the synchronic action both of catastrophic events together with slow or gradual processes. Among the first ones are included the sliding fluxes originated by turbidites, the meteoritic impacts, the volcanic explosions and the earthquakes (see Hsü 1993). And referring to the sudden biological changes, there are mentioned among others, the chromosomal mutations such as polyploidy and mutations in the regulating genes that take part in embryogenesis. On the other hand, it is possible to admit the existence of very slow changes both in the terrestrial crust as well as in the organisms populations. The last ones can be detected, for example, when the erosive agents modify very slowly a landscape of geological stable zones. However, microevolutive modifications can be expected in large populations living without strong pressure of selection. For the latter slow changes a strict saltationist biologist could warn us, rightfully, that with the use of small analysis scale, the changes are discontinuously registered. From an opposite point of view a Darwinian paleobiologist could interprets the gaps (for example those of fossil record) as failures or shadows due to the disproportion between the time-scale represented in a few meters of sedimentary rocks outcropping. In this frame, the conventional biostratigraphical approach could be insufficient in order to recognize lacunas in the fossil record. An uplifted vision of Uniformitarianism, postulates in Geological past-time the performance of natural processes, even though the active agents remain infrequent or attenuated.

Simpson demonstrates the limitations (or the impoverishment) of the controversy: continuous against punctuated evolutive phenomena. In the development of Chapter III: "Micro-Evolution, Macro-Evolution and Mega-Evolution", Simpson (1944) advances and solves the futile discussion that will take place towards the 80-90's between the denominated punctuated evolution and gradualist change. The same author emphasizes the coexistence, in fossil register, of cases of quantum evolution (taquitelic), normal (horotelic) or slow (braditelic). It seems interesting to transcribe from Simpson (see preface for 1984 edition of *Tempo and Mode*, p. xxvi) when discussing the differences between the Neo-Darwinist notion and the punctuated evolution by Gould, he writes down:

> We thus have here an "either-or" proposition a Hegelian o Marxist Dialectic. It is apt to point that apparent contradiction between thesis and antithesis leads logically not one to the other but to synthesis. That is what the synthetic theory does.

Nevertheless, in the same work Simpson adopts a conservative point of view when stating that the evolutionary mechanisms responsible for the cases, both of gradual or punctuated evolution, were the same. In this way, his reflections rather than contributing with overcoming explanations, seem to keep the Darwinian proposal within the same questionable zone. The contributions of Simpson, instead of explanations "*ad hoc*", lead to a disquieting zone for part of the threatened Darwinist proposal.

3.2. The notion of species within the framework of fossil record

In spite of the effort made by numerous generations of biologists and paleontologists from different schools, the notion of species as real entities of nature remains a conflict without solution for the Theoretical Biology. The state of present confusion has been "systematized" by Wilkins (1997). This author summarizes in its Table 1, three levels of distinction, with eight nodes and their corresponding polar definitions. The first level of distinction is the one that discriminates the instrumental notion of species from the other realistic ones.

The definition of the term species formulated by Darwin (1959) is very simple (see below 4.1). However, each species is considered by Darwin, as well as by many Darwinists, a tangible entity. If it weren't like this, then it would not have an origin and it could not transform. In section 4 it will be explored with some detail the reasons for this difference between an instrumental definition, above mentioned, and the realist conception.

Regarding *the species problem in paleontology*, Simpson (1961) proposes a "Solomonian-solution" in order to maintain the classic notion of species as real entities of the nature through time. Good examples of evolutionary lineages such as those discussed by Simpson have been shown by Bettenstaedt (1962, 1968). The latter studied allochronic (successive) populations of benthonic foraminifers with a low evolutionary rate (cases of microevolution). The analyzed populations were numerous enough to exhibit statistically representative designs. The samples came from marine Cetaceous sediments from the Niedersaxen Basin, following each other continuously in time. Among the different cases studied by the same author, there are examples of transformation of lineages which do not branch out ("*Populationsumwandlung*" + "*Artumwandlung*"). Such is the case of *Vaginulina procera* (in Bettenstaedt 1962, figs. 4-5, 1968, fig. 4). There are also different examples of cladogenesis ("*Abspaltung*") exhibiting branches of differentiated populations with new species appearing. Such is the case of *Lagena hauteriviana* s.l. (in Bettenstaedt 1962, fig. 13, 1968, fig. 8). Representative illustrations for two cases studied by Bettenstaedt are shown in figures 1 and 2 of the present contribution. These and other similar sequential lineages show the conflict that Simpson tries to solve with the aid of optional cuts within the respective lineages (See fig. 3, tacked from Simpson). This arbitrary procedure would allow segregation of different morphologic species, within the continuous evolution for cases of microevolution. The species definition is consequently modified by the author (Simpson 1961, p. 153) as follow:

> An evolutionary species is a lineage (an ancestral-descendent sequence of populations) evolving separately from other and with its own unitary evolutionary role and tendencies.

The same author will maintain, by means of an *ad hoc* artifice (Simpson 1961, figs. 13 B and C'-C" and 14), the use of the taxonomic category (see figure 3). This is perhaps a strong argument suggesting that species as a real organization in nature, lacks an identity in the course of evolution.

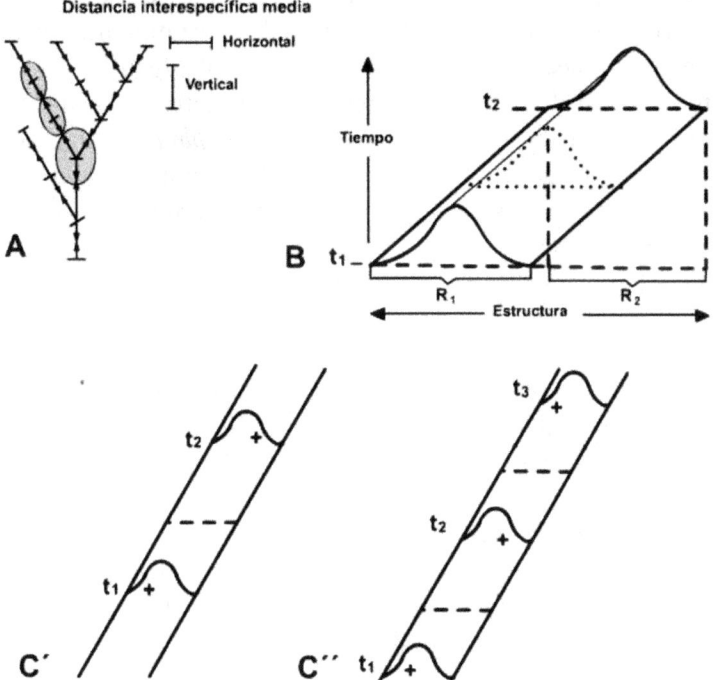

Figure 3. The *"problem of the Species in Paleontology"* according to Simpson (1961). This author solves it in a *"Solomonian"* way, slanting arbitrarily the lineages.

3.3. The sense of change and the progressive reduction of variability in the evolution

Darwin admitted the possibility of establishing guidelines or *rules taking care of growth and phylogeny*. This author considered in "Developing and Embryology" (Darwin 1989, Chapter XIV) previous proposals of von Baer (1792-1876) and Haeckel (1834-1919) according to which ontogeny exhibits stages that correspond with ancestors from the respective living groups. In his opinion, the variations in the numerous descendants of a remote ancestor have appeared in a not very early period of their life and they are inherited in a corresponding period. However, the same author does not seem highly disposed to examining and weighing up the relevance of these evidences without sharing those linked to the external environmental evolutionary aspects as a determinant of the evolution of organisms. It is interesting to mention that an acceptance without restrictions of the Recapitulation-Law saying that "*ontogeny recapitulates phylogeny*" has been disqualified by the same Haeckel as Law of general compliment. The same author recognizes different cases of exceptions that were included within heterochrony.

A modality in the directionality of the change of the organisms constitutes cases of hypertelic growth in where an out of proportion increase in a structure within the organism is verified. Cases of the paleontological record, exhibit a cumulative increase of this allometric growth of body components through the time within a particular lineage. Different authors have suggested these evolutionary designs can better be understood accepting a necessary interdependence between the biological process and its determining bio-molecular agents. After *The Origin of Species*, paleontological Literature collects throughout decades a conclusive data on the existence of directional evolutionary tendencies shown by fossil record.

This relation does not have an appropriate place within the Darwinist vision of the change. According to this, Simpson (1953, p. 245) admits that:

> Since the evolution is to some evidently large degree nonrandom and oriented. phylogenetic sequences or many of them have some element of sustained direction; they show a prevailing tendency or in other words they have trends

and after:

> Almost all fossil sequences long enough to be called sustained show prevailing tendencies in some characters and over part, at least, of the sequence.

Does Simpson accept then the presence of endogenous factors that actively participate in the directional modality of change? The answer must be negative. The *ad hoc* explanation that the same author proposes to interpret the causal factors that explain different evolutionary tendencies from fossil registry have nothing to do with the presence of endogenous directional tendencies that imprint attributes that stand out in the successive clades. Let us examine, for example, the interpretation that Simpson provides to the case of the iterative evolution of the Ollenilids, a group of trilobites from the Cambrian Age studied by Kaufmann (see Simpson 1953, p. 248, fig. 31). Based on the fossil registry of Sweden, Kaufmann recognized the appearance of four successive lineages of the *Ollenelus*-type, each one of these constituted by forms that evolve according to similar tendencies in the different lineages. The Iterative Design represents the branches that arise successively from an unknown ancestor unknown of the region and that would have to hold a conservative character regarding the permanence of characters in the rear part of skeletal remains (pigydium) of these arthropods. To Simpson, the reiteration of lineages would reflect the recurrence in the environmental conditions that exert selection pressure on conservative ancestors consecutively arriving to this biotope. In his interpretation, these successive lineages (Simpson 1953, p. 249):

> apparently represent times of local environments unfavorable to these animals with resulting local extinctions of older lines followed by re-invasion when the environment again became favorable.

This example is representative of a collection of cases taken from different fossil groups. Such could be the cases of the torsion of umbo in the plane of valve-contact exhibited by several clades of the extinct bivalve *Exogyra*, or the case of the winding plane in *Gryphaea* (*sic*) or in the weakly opistogyre valves of the bivalve *Ostrea*, among others (Simpson 1953, p. 248). The mentioned example is a case of an ample and diverse set of evolutionary tendencies (trends). A detailed examination of the different evolutionary tendencies that reflects the fossil registry and the problems within each one of these, exceeds the objectives of the present contribution (additional Information on evolutionary tendencies can be found in Osborn 1934, who introduces the term *aristogenesis* for *orthogenesis*, Beurlen 1937, mainly 4, pp. 66-73, Rensch 1959, Hallam 1977, or Valentine 1985, among other authors).

Within section 3.3. it now seems appropriate to consider the ideas of Rosa (1900), and Fechner (1873) (*fide* Hennig 1968, pp. 298-299) dealing with the *progressive narrowing of variability*. Beurlen (1937, p. 80) clarifies the subject in the following terms:

> die Evolutionsbreite der aufeinanderfolgenden Gruppen eine deutliche Einengung, da die grundlegenden Organisations divergenzen immer geringer werden: der Typus der Säugertiere ist einheitlicher und geschlossener als der Reptilien, der sinerseits wieder als unbedingt einheitlich erscheint gegenüber dem Typus der Amphibien-Stegocephalen [...]. Die Gleiche Erscheinung wiederholt sich in jeder systematischen Einheit höherer Order niedrigerer Ordnung. (see Hennig 1968, p. 300)

It is well-known the case of irruption of marine life in the fossil record during the interval Vendian-Cambrian, by the end of the Proterozoic – beginnings of the Fanerozoic, approximately 600 Ma ago (see Valentine 1986). During this geologic time interval, which is relatively short comparing it to the Phanerozoic Era, appear almost all Phylums that are known in the fossil record. After the Cambrian, new structural types of similar ranks will not arise. The previous case is mentioned as an example of the phenomenon of macroevolution (or megaevolution). Analogous examples, although implying taxa of smaller rank, are known for the orders of mammals towards the beginning the Cenozoic, the Archosauria (diapsid reptiles) in the beginning of the Jurassic, the pisciforms towards the Devonian Period or, finally, the Divisions of the Vegetal Kingdom that are registered almost completely during the interval Late Silurian-Devonian. After the irradiation of anatomic-functional novelties, new structural types within a level of equivalent organization (Phylums or Divisions), will be limited. In time each group will unfold following its own tracks in order to keep basic biological characters. For example, within the corals, evolutionary trends do not overlap the space of annelids, or ferns do not evolve towards lycophytes.

Valentine (1989) recognizes two main evolutionary designs within the fossil record. The first one, involves the **levels** of organization (= *anagenesis* according to Valentine 1989) represented by the arising of new structural features as novelties. The second, or evolution by **clades** (*cladogenesis* of the gr. = branches), depict the spread of systematic units, hierarchically subordinate to other ancestral ones, from which they branch like lateral branches from a main stem. The Valentine's design of evolution by clados roughly depicts the frame in which the reduction of variability deals, as stated by Fechner-Rosa. This restriction takes place after an irradiation event. However, the same does not give account of the appearance of evolutionary new features, involving grades of increasing complexity, which assure the freedom of the evolutionary process.

Is it possible to explain the direction as well as the progressive restriction of diversity, keeping a neutral behavior of the endogenous evolutionary factors? In other words, do genes behave passively prior to the action of natural selection? At present time there are crescents evidences of the role of immunological systems as sensors of environmental change as well as the response provided by the organism, including the manipulation that enzymes of the genetic message exert as a reply to the signals of the environment. There is also an increasing knowledge on the role of the regulating genes in morphogenesis. Finally, the importance of paedomorphosis is more and more emphasized; thus, generating new features, previously anticipated in the ontogeny. These possibilities, that enrich the evolutionary theory, are outside the subject of the present discussion that is essentially historical.

3.4. The notion of increasing complexity

The *increasing complexity* shown in the fossil record, deals with the successive appearance of groups with morphologic-functional structure more and more elaborate or specialized. The outline of five kingdoms of the living organisms according to Whitaker (1980) is ordered according three main steps: first the prokaryotes, then the protists, being located in the last ones, the animals, plants and fungi. This ascending order, in which the simple organisms underlie the complexe ones, corresponds with the order of appearance of such groups in the geologic time. Additionally, it must be pointed that new groups follow one another without diminishing their diversity, missing or moving the previous simplest ones. Cases of evolution by degree, like the latter one, are numerous in the fossil registry, although the category of these latter groups is smaller than those of the previously mentioned example dealing with kingdoms. Amongst the plants, for example, those with seeds go after those reproducing by spores and those having flowers (the most complex structures in the reproduction) appearing lattermost. Similarly, amongst the vertebrates groups with elaborated nervous or circulatory systems pursue the simplest one. This evolution by degree has been denominated "Anagenesis" by Rensch (1959, p. 289ss.; see also Szathmáry & Maynard-Smith 1995).

The status of the increasing complexity notion, within Darwinism, does not seem clear:

> The embryo in the course of its development, rises generally in organization; I use that expression although I confess that it is almost impossible to define clearly what means when speaking of an organization that is superior or inferior. (Darwin 1869, p. 555)

It seemed appropriate to avoid a possible fifth difficulty dealing with the restrictions of the Malthusian perspective. According to Darwin's words, the reading of Malthusian thesis assisted him to fine-tune his ideas on the way in which natural selection operates favoring the most apt to survive. This assumption was favorably accepted by the Society, where the theory was born and settled. To this circumstance, it should be of interest to mention the defense assumed by sociobiologists of the role of cooperation (in terms of symbiosis, commensalisms, or partition of work, among other modalities) in the evolution. Arguing for their assumptions an orthodox-Darwinian position, they assert natural selection operates actively, enhancing groups or individuals, both cases willing to cooperate (see Haldane 1932, Clutter-Broch & Harvey 1978, Wilson 1975, 1998).

4. The philosophical substratum of currently accepted proposals of the theory of biological evolution and the conflictive parts

In sections 2 and 3, an approach was procured in order to recognize the currently accepted aspects of the Theory of Evolution and the conflicts that rise when the Darwinist vision of change must be concurred with fossil record. The present Section deals with the philosophical "anchorage" of the "stable" theoretical statements as well as the others conflictive assumptions currently in debate. The focus is the search for notions that naturalists adopted as basic terms. The thesis is the possible role of cultural background in the process of reasoning and the supply of concepts as well as primary terms. This source provides postulates that scientist accepted, perhaps inadvertently, as premises within a "modern" perspective for natural sciences. In this frame, this contribution supports the point of view according to which permanent achievements of Darwinism are in correspondence with the renewed philosophical thought of the XVIIIth century. Among these are mentioned: the population analysis, the natural selection and the meaning of the fossil registry. However, some conflicts suggest an anchorage in "ancient" philosophy, present in the cultural formation of the Darwinian naturalists: the dialectic thought transferred to the observable facts (antinomies without solution), the pressures exerted by the notion of teleology, and the notion of species, adopted by the theory as a primary term.

4.1. The stable parts

The phenomenon of change within the living beings is a part of the never-ending transformation that exhibits the tangible world and that has been perceived in different cultures. The elaboration of an evolutionary theory was a merit of the XIXth century biologists. The principle on which it is based allows it to assimilate the experience of population analysis, revealing of the discontinuous nature of the diversity in space, as in time, this last one shown by the bio-stratigraphic record. The definition of the states of characters in any taxon is the result of the average of the different values measured in individuals belonging to the corresponding populations. It is senseless to investigate the similarity or distance to an ideal archetype, which exists only in the representation of the taxonomist.

The study of the variability as well as the probabilistic processes and their calculations has an important record in the XVIIIth Century. Particularly, the work of Bayes (1702-1761), including his theory of decisions, stands out with increasing importance in present philosophy of science. It is true that Darwin has not made use of the Conditional Probability, Bayes' Theorem. The mention of late author wishes only to emphasize the influence that at that time, had the reflective vision of man about the phenomena of diversity and change. From now on Philosophers and Mathematicians will pay a renewed attention to the changing world of sensible phenomena. These will not be considered a tricky kaleidoscope of "shadows" that must be left aside in order to take the path towards knowledge as Parmenides and Plato considered. The new philosophical vision will be one that "waking up from the Metaphysical dream", influenced by different authors like Hume (1711-1776):

> But this obscurity in the profound and abstract philosophy, is objected to, not only as painful and fatiguing, but as the inevitable source of uncertainty and error. (Hume 1739-1740, I, p. 4)

Hume (1739-1740) states that the relation of cause and effect does not constitute an axiom. This relation arises from daily experience, which does not allow generalizations either.

> There are no ideas, which occur in metaphysics, more obscure and uncertain, than those of power, force, energy or necessary connection, of which it is every moment necessary for us to treat in all our disquisitions. We shall, therefore, endeavor, in this section, to fix, if possible, the precise meaning of these terms, and thereby remove some part of that obscurity, which is so much complained of in this species of philosophy. (Hume 1739-1740, VII, p. 28)

A consequence of the previous statement is the critic to the value of induction as a truth criterion:

> I shall content myself, in this section, with an easy task, and shall pretend only to give a negative answer to the question here proposed. I say then, that, even after we have experience of the operations of cause and effect, our conclusions from that experience are not founded on reasoning, or any process of the understanding. This answer we must endeavor both to explain and to defend. (Hume 1739-1740, II, p. 14)

The species according to Darwin. The exposition that this author does on subject of the species in Biology allows a dual interpretation: together with an instrumental presentation for the term species, a realistic notion of the unit is accounted. This duality will be analyzed below. First the simple definition of the term by Darwin (1859):

> I look at the term species as one arbitrarily given for the sake of convenience to a set of individuals closely resembling each other [...] it does not essentially differ from the term variety which is given to less distinct and more fluctuating forms.

This definition, which could be considered as instrumental, lacks, essentialist and / or typological implications. In addition, it could correspond with the skeptic vision of Hume (1776):

> The academics always talk of doubt and suspense of judgment, of danger in hasty determinations, of confining to very narrow bounds the enquiries of the understanding, and of renouncing all speculations which lie not within the limits of common life and practice. (Hume 1739-1740, V, I, p. 18)

4.2. Focusing on the conflicts

In spite of the definition quoted above, specie is considered by Darwin and many Darwinians (in particular neo-Darwinians) as a real entity within nature. If it weren't like this then it could not have an origin and it could not become transformed. What is the reason for this difference between an instrumental definition[1] and a realistic conception?

In this contribution the following statement is proposed: Two premises of Darwinism, the notion of species and the hierarchic system, mark the presence of traditions coming from Ancient Philosophy, which have been incorporated by the naturalists, presumably, in its basic formation. The taxonomic categories respectively arranged in increasing order, correspond with the *primary substance* and *secondary substance* notions, both proposed by Aristotle in the *Organon*.

Next, an approach to the *Categories* will be attempt, where Aristotle makes a systematic treatment on *"which he is said of"* or *"it is predicated of"*. First the *substance (ousía)*. Here an early difficulty should arise for some authors: in spite of occupying the place of first category (within the predicables) the substance seems, however, after some to correspond to the *being* or the entity (Miguel Candel Sanmartín 1982, p. 35). However, *substance* is that *"which he is said of"* (Miguel Candel Sanmartín 1982, pp. 7-8) and different hierarchic ranks fit to it:

Primary substance: it is what one does not affirm of a subject and what not a subject is. (see: *Categories*).

---

[1] The species, as an instrumental unit, will be supported during second half of the twentieth century by the Numerical Taxonomy discipline, using "operative" approach techniques. The species, and also the remaining systematic categories, will became Taxonomic Operative Units (TOU) in order to build a data matrix. The species displayed in this arrangement adjust without difficulty, to Darwin's definition (1856). It is left outside of this contribution the treatment of epistemological difficulties for "operationalism".

Secondary substance: what it is said of the species in which the primary substances are embraced. (see: *Categories*)

The term *Substance* (from the Latin: *substantia*; see Ferrater Mora 1979) seems to allude to what underlies, what sustains, what gives permanence in spite of the changes and accidents that happen in the individual subject. Would Aristotle accept to consider growth as an accident underlying the substance of the Being? It seems appropriate to remark that the notion of *semaphoront* by Hennig (see Hennig 1968, p. 8: "the individual one, as it appears during a very restricted lapse of its existence") seems to be in correspondence with the proposal by Aristotle of *Smallest Substance* (= individual).

According to Aristotle, *species* (*eidos*) is more substance than genus (*genos*). In addition, the term genus is broad including categories of diverse rank (in current Systematic Biology, the Aristotelian term *genus* embraces all the categories of rank higher than species: genera, families, order, classes and phyla). The same author admits the following relationships between categories: *primary species* (individuals as: "*this man*" or "*this horse*") < *secondary species* (animals), were *secondary species* include several possible ranks: man (or horses) < animals.

The "encapsulating" hierarchic system (or a "Chinese box" like system: a box that once opens contains another box and so forth), is proposed by Aristotle as follows:

> In cases of subordinated genus, those of higher rank are predicated by those of lower rank, in such a way that all the differences in the predicate are also differences of the subject (see Categories: 3)

That is:

	Primary substance	Secondary substance
Sócrates	< man	< animal
Minimal species	< species	< genus
	(*eidos*)	< (*genos*)

The presence of the Aristotle's thought is manifested in contributions of philosophers of Post-Socratic and medieval tradition. Such is the case of Porphyries (see Eco 1983, figs. 2.5 and 2.7), who proposed his well-known tree (*Arbour Porfhyriana*). The latter is a "model" for the hierarchic system adopted later by biologists, both pre-evolutionists as well as Darwinians. Each category fists in a more general one (except for "the widest genus"). These ones, as well, include the most special species until we reach at the "smallest species". An always present care in Aristotle, as well as in Post-Socratic and the medieval tradition devoted to a detailed analysis of Categories (see Spade 1996), has been the problem of definition (what is to define, or *definiendum*, and the expression that does it, or *definiens*). When should be it defined by differences?, and when by properties? Eco argues that Porphyrius's Tree is only constructed by differences: "Genera and species are composed only of differentia" (Eco 1990, 2.2: Critique of the Porphyrius's Tree). This last one is the criterion followed by taxonomic procedures in Systematic Biology to establish diagnosis and differential diagnosis of taxa (see Mayr and Ashlock 1991, pp. 413-414):

> Differential diagnosis: A formal statement of the characters that distinguish a given taxon from other specifically mentioned equivalent taxa.
> Diagnosis: In taxonomy, a formal statement of the characters (or most important characters) that distinguish a taxon from other similar or closely related coordinate taxa.

The Aristotelian notion of limit or interval (see Ferrater Mora 1979, p. 731; also appears in Mayr & Ashlock 1991).

> Description: In taxonomy a more o less complete formal statement of the characters of a taxon without special emphasis on those which set limits to the taxon or distinguish it from coordinate taxa.
> Delimitation: In taxonomy, a formal statements of characters of a taxon which sets its limit.

In Biology it is common to admit that systematic categories of higher ranks than species do not have correspondence with real entities of nature. However, in Phylogenetic Systematic (see Hennig 1968) or cladism, each branch originated by lateral loosening in the evolutionary process occupies a place in the hierarchic system whose organization corresponds to the process of cladogenesis (in this discipline, the hierarchic system is able to reflect phylogeny). It is beyond the purpose of the present paper to examine the discussion of the cladism assertions, which could be described thesis with realistic attributes in this concern. Resembling Aristotle (see above: *"the species is more a substance than the genus"*), the biological species is considered by many

Darwinian biologists also an entity whose existence within nature is more "real" than the genus. Finally, some procedures of traditional systematic could be related to the Aristotelian inheritance, (the cladism strategy proposed by Hennig must be excluded of this assertion). In Aristotle, the genera (i.e. in plants or animals) gather species sharing attributes. Species are settled down, however, by distinction.

The essentialist definition of species in pre-evolutionary biology. The conflicts displayed by the notion of typological species due to the influence of platonic conception have been properly solved by Darwinism (a term conflict must be stated first: in many quotes, *genon* in Plato is equivalent to *eidos* in Aristotle). A valuable analysis of this subject has been exposed by Mayr (see "*The Typological Species Concept*", in Mayr 1963, pp. 16-17). When Darwinism incorporates the population analysis in its working methodology, it can early detach from the archetype-notion. In a similar way, Darwinism was not affected by the scorn that change phenomenon produced in Parmenides and Plato's philosophies. The notion of change/corruption mixed with that of permanent change is faced with that of essence/idea. The later antinomy, which relates to anti-Heraclites extremism, was decisive in the cultural formation of pre-evolutionists biologists.

However, as it was pointed out before, Darwinism dragged the acceptance of notion of species as a real entity in nature, whose origin was to be explained. The cause of this dual proposal of Darwinism (instrumental definition / realistic notion) is beyond the aims of the present contribution. Perhaps the adoption of the notions of species and the hierarchic system would have to be tracked down to a sociological investigation rather than an epistemological one.

What is difficult to understand in the historical debates about Darwinist theory, is the silence about the contributions arrived in the framework of the Middle Ages Philosophy about the subjects: particular substances / universals as well as on the related topics: realism / conceptualism / nominalism. This historical background (see Gilson 1965) and the proposed solutions, particularly during XIIIth Century, allow considering a part of the present debate on the species in biology repetitive and irrelevant. In such sense, the point of view of Pierre Abelard seems enlightening; according to this author, the "universals" (*res universalis*), although are not only terms devoid of meaning (or *flatus vocis*), also they are not objects or perceptible "substances". After Abelard, these universal ones are only confused images, lacking of existence in the nature; the same turn out from the predicament of many individuals, those that yes constitute however real ontological subjects.

The opposite: gradual / punctuated process. Historical debates on theories, that took place in the last two centuries on theories dealing with the Geology and the Paleontology, are useful when an approach to the scientific thought characterizing the respective scientific communities is procured. The dialectic framework of the controversy about the evolutionary factors in evolution at the beginning of the 20th century is of interest to the subject. This conflict between natural selection (external factors) vs. mutations (endogenous factors) will show an end of sort "synthesis of the opposites".

However, in the last century, the case dealing with continuous vs. punctuated processes (see 3.a. above) does not give notice for a similar end. This debate is strongly related to the catastrophism vs. uniformism that took place during the 19th Century and was renewed towards the end of the last Century in the frame of the Sequential Stratigraphy Theory. The struggle between opposite *gradual vs. discrete* seems to form part of a comprehensive human vision, based on philosophical thinking, which is projected over the phenomenal world embracing, as in our case, factual scientific disciplines. As it was previously emphasized, the bioestratigrafic record exhibits the joined presence of lineages showing continuous evolutionary-designs, together with other punctuated ones, both belonging to different groups keeping their own evolutionary rates. The differences in the evolutionary rate mentioned, can be explained by taking into account that several unrelated evolutionary factors operate on the evolutionary processes. The latter suggests that the change should be then measured in terms of a polynomial equation. Taking into account this data, the conflict between opposites: gradualism *vs.* punctuated equilibrium during the neo-Darwinian period should be restricted to dialectic theoretical portray.

Trends in evolution. Debates related with directional evolutionary trends in lineages of the fossil record, can be better understood focusing the significance attributed to the words teleology and teleonomy. Let us consider, at first, some involvements of the term teleology.

Teleological explanations like "*Why the human beings have lungs?*" entail the answer concerning the need of oxygen for the combustion of nutritional substances in the body. The last example indicates the existence of a purpose or intentionality. The latter is similar to the task in a furniture factory, for example, guiding the manufacture of chairs by seats. The Theory of Finalism in paleobiology is related to the Aristotelian notion of final causes that implies, from the beginning, a determination in the search of an advanced objective.

In the middle of the 20th century, Neo-Darwinist biologists, who admit "evolutionary inertia" (see Simpson 1944), will try explanations sustained in the external environmental control. Thus, the progressive size-increasing in fossil trends, is explained as the result of the environmental pressures ruling the fitness. This way, a careful distance from supposed metaphysical involvements related with causality and teleology is kept. In particular, this is evident by the reluctance to admit possible "lethal evolutionary trends" in the fossil record, the later ones entirely ruled by endogenous factors. The argument that Simpson offers in Chapter II, titled: "The Fundamental Problem, in relation to the forces that acted during the history of the life" (Simpson 1951) enlighten this outlook. This Chapter begins with a dialectic approach:

> are these the same forces that build on the rest of the material universe...? (materialistic, mechanicists or causalists). Or are they characteristic forces of life and inherent to it? (vitalistic).

Soon, Simpson recognizes a third possibility:

> Or do they include principles that transcend matter (substance) and life itself...? (finalists).

However, this third possibility is a slight variant of second one above:

> And this last one has been considered as one of the innumerable variations of vitalism.

Finally, Simpson's interpretation according to which trends in the fossil record must be described within the framework of the Aristotelian notion of teleology:

> The distinctive belief of finalists is based on direction towards a goal or an aim. According to them this aim is not attainable due to what it precedes it, but what it precedes. It is only a means to obtain the aim. The goal, in spite of appearing later in time is the cause, and the course of history that precedes it is the effect.

Distant from the finalist perspective, some paleobiologists believe that directionality in the fossil record is guided in many cases by teleonomic processes (see above 3.3.). Regulatory device, present in the ontogeny, became projected in their controlling process through successive evolutionary lineages. The "behavior" of a tendency can better be interpreted with the availability of a program (or a code of information), but does not entail a principle of causal end. In other words, the process is not guided by an intention of the sort: *"the birds have acquired their wings to fly"*. This way, the evolutionary tendencies of the fossil record do not follow a course delimited by the environmental pressures. Two explanatory theoretical models, coming from Biology, allow noticing differences in the meaning of the terms teleonomy and teleology. In the first place the teleonomic outlook for the embryological development of organisms (embryogenesis) is mentioned. This process, responds to a pre-established informative structure (the genetic code), which is own for each class of the characterized organisms to share a common pool of genes. The second example deals with the strategy used by the *Escherichia coli* bacteria (that lives, among other parts, in the human intestine) to fit its enzyme production. This procedure, has allowed to Jacob and Monod the construction of the Operon model (Monod 1961) which possibilities to explain the phenomenon of the biological adaptation of organisms on the basis of the necessary interdependences with the biomolecular-device controlling the process. The pre-established informative structure (or, if it is preferred, the program for the respective genetic codification) has not been constructed *"ad hoc"* because it does not exist a final cause previously planned.

The term *teleonomy* is only analogous to the term *teleology*. However, the significance of the former keeps a deep disconnection with this second concept bounding with the purpose notion. The expression "determined natural teleology" (Dobzhansky *et al.*, pp. 497-498) seems similar to that of teleonomy. However, the meaning of the included word teleology could bring confusion because it bears on the intentionality notion and its connection with "super-natural" powers. To this last ones concerns the necessity for an overseer driving the changes of organisms, dealing the increasing complexity in the levels of anatomical-functional organization through the time and supporting the irreversibility of the biological evolutionary change. In this way, unsolved or dodged problems, such as the teleonomy and the increasing-complexity, are presumably related to the reluctance in adopting supposedly metaphysical assumptions.

5. Conclusions

Some specific contributions of the Darwinism, confirmed as stable components of the evolution theory, seem to be in correspondence with the renewed vision for sciences achieved by Modern Philosophy. These contributions deal with the meaning of the fossil record, the population analysis and, finally, with the dependent theory of natural selection as evolutionary factor.

On the other hand, some concepts admitted or adopted as premises in the Darwinian theoretical vision of the change, exhibit a link with Ancient Philosophy. Presumably, this heritage was acquired in the basic education received by the naturalists. Such is the case of the conflictive category of species, coming from the Aristotelian philosophy. However, other unsolved or dodged problems, such as the teleonomy and the increasing-complexity, are presumably related to the reluctance in adopting supposedly metaphysical assumptions.

The biostratigraphic record confirms the joined presence of lineages showing continuous evolutive-designs together with punctuated ones, both belonging to different groups having their own evolutionary rates. Taking into account this data, the conflict between opposites: gradualism *vs.* punctuated equilibrium seems dialectic theoretical portray, characterizing debates during the neo-Darwinian period.

Several and unrelated evolutionary factors guide the evolutionary processes. The change should be then measured in terms of a polynomial equation.

The difficulties recognized in the historical Darwinian Theory, do not diminish the role of its contribution as the most relevant attempt to account for a comprehensive picture of the permanent change in living beings.

References

Ager, D. (1993), *The New Catastrophism. The Importance of the Rare Events in Geological History*, Cambridge: Cambridge University Press.

Aristóteles (1946), *Organon. I. Catégories*, Paris: Vrin (French version by J. Tricot).

Aristóteles (1966), *Aristotle's Metaphysics*; Bloomington: Indiana University Press. Stanford Encyclopaedia of Philosophy, 2001, available in: <http://stanford.edu/aristotle>.

Aristóteles (1982), *Tratado de Lógica (Organon) I: Categorías-Tópicos-Sobre las refutaciones sofísticas* (Spanish translation by Miguel C. Sanmartín), Madrid: Biblioteca Clásica Gredos.

Ayala, F.J. & T. Dobzhansky (eds.) (1974), *Estudios sobre la filosofía de la biología*, Barcelona: Ariel.

Bettensetaedt, F. (1958), "Phylogenetische Beobachtungen und der Mikropaläontologie", *Paläontologische Zeitschrift* 32 (3/4): 115-140.

Bettenstaedt, F. (1962), "Evolutionsforgänge bei fossilien Foraminiferen", *Mitteilungen – Geologisches Staatsinstitut in Hamburg (Sonderdruck)* 31: 385-460.

Bettenstaedt, F. (1968), "Wechselbeziengungen zwischen angewandter Mikropaläontologie und Evolutionforschung", *Beihefte zu den Berichten der Naturhistorischen Gesellschaft zu Hannover* 5: 337-391.

Beurlen, K. (1937), *Die stammesgeschichtlichen Grundlagen der Abstammungslehre*, Jena: Gustav Fischer.

Clutter-Broch, T.H. & P.H. Harvey (eds.) (1978), *Readings in Sociobiology*, San Francisco: Freeman & Co.

Cuvier, G. (1817), *Discourse on the Revolutionary Upheavals on the Surface of the Globe and on the Changes which they have produced*, Paris: Dufour et D'Ocagne.

Darwin, C. (1859), *On the Origin of Species by Means of Natural Selection or The Preservation of Favoured Races in the Struggle for Life*, London: John Murray.

Díez, J.A. & P. Lorenzano (eds.) (2002), *Desarrollos actuales de la metateoría estructuralista: problemas y discusiones*, Bernal: Universidad Nacional de Quilmes.

Dobzhansky, T. (1966), *La evolución la genética y el hombre*, Buenos Aires: Eudeba.

Dobzhansky, T., Ayala, F.J., Stebbings, G.L. & J.W. Valentine (1980), *Evolución*, Barcelona: Omega.

Eco, U. (1983), *Semiotics and the Philosophy of Language*, Bloomington: Indiana University Press.

Ferrater Mora, J. (1979), *Diccionario de Filosofía*, Vols. I-IV, Barcelona: Atlántida.

Gilson, E. (1952 [1965]), *La filosofía en la Edad Media*, Madrid: Gredos.

Grande, L. (1989), "Vicariance Biogeography", in Briggs, D.E.G. & P.R. Crowther (eds.), *Palaeobiology (A Synthesis)*, London: Blackwell, pp. 448-451.

Haldane, J.B.S. (1932), *The Causes of Evolution*, London: Longnams Green.

Hallam, A. (1977), *Patterns of Evolution as Illustrated by the Fossil Record*, London: Elsevier.

Hennig, W. (1968 [1950, 1966]), *Grundzüge einer Theorie der phylogenetischen Systematik*, Berlin: Deutscher Zentralverlag. (English translation: *Phylogenetic Systematics*, Urbana: University of Illinois Press, 1966. Spanish translation: *Elementos de una sistemática filogenética*, Buenos Aires: Eudeba, 1968.)

Holmes, A. (1952), *Geología física*, Madrid: Omega.

Holmes, A. & D.L. Holmes (1980), *Geología física*, Madrid: Omega.

Howell, B.J. (1962), *Introducción a la Geofísica*, Madrid: Omega.

Hull, D.L. (1984), "Cladistic Theory: Hypothesis that Blur and Grow", in Duncan, T. & T.F. Stuessy (eds.), *Cladistics: Perspectives of Evolutionary History*, New York: Columbia University Press, pp. 5-23.

Hume, D. (1776 [1910]), *An Enquiry Concerning Human Understanding*, Harvard Classics, Vol. 37, New York: Collier and Sons.

Hsü, K. J. (1983), "Actualistic Catastrophism", *Sedimentology* 30: 3-9.

Lyell, C. (1830-1833 [1990]), *Principles of Geology*, Chicago: The University of Chicago Press.

Mayr, E. (1963), *Animal Species and Evolution*, Cambridge, MA: Belknap Press of the Harvard University Press.

Mayr, E. (1982), *The Growth of Biological Thought. Diversity, evolution, and Inheritance*, Cambridge, MA: Harvard University Press.

Mayr, E. (1988), *Toward a New Philosophy of Biology*, Cambridge, MA: Harvard University Press.

Mayr, E. & P.D. Ashlock, (1991), *Principles of Systematic Zoology*, New York: McGraw-Hill.

Musacchio, E.A. (2001), "Procesos evolutivos comparados en disciplinas fácticas: isomorfismos o interdependencias necesarias?", *Episteme* 12: 47-59.

Musacchio, E.A., (2003), "Cambios Geo-históricos y Registro Fósil: teorías comparadas", in *III Simposio Internacional Principia, Resumos*, Florianópolis: UFSC, pp. 55-57.

Musacchio, E.A. (2004), "Procesos recurrentes y procesos irreversibles en geología histórica", in Martins, R.A., Martins, L.A.-C.P., Silva, C.C. & J.M.H. Ferreira (eds.), *Filosofia e história da ciência no Cone Sul: 3º Encontro*, Campinas: AFHIC, pp. 144-152.

Osborn, H.F. (1934), "Aristogenesis, the Creative Principle in the Origin of Species", *American Naturalist* 68: 193-235.

Owen, H.G. (1976), "Continents Desplacements and Expansion of the Earth During the Mesozoic and Cenozoic", *Philosophical Transactions of the Royal Society of London* 281: 223-291.

Popper, K. (1934 [1959]), *The Logic of Scientific Discovery*, London: Hutchinson & Co.

Popper, K. (1963), "Truth, Rationality, and the Growth of Scientific Knowledge", in Popper, K., *Conjectures and Refutations*, London: Routledge & Kegan Paul, pp. 215-250.

Rensch, B. (1959), *Evolution Above the Species Level*, London: Methven and Co.

Rosa, D. (1900), "La réduction progressive de la variaibilité et ses rapports avec l'extintion et avec l'origine des spèces", *Archives Italiennes de Biologie* 33: 314-318.

Rosa, D. (1931), *L'Ologénèse. Nouvelle Théorie de l'évolution et la distribution géographique des êtres vivants*, Paris: Félix Alcan.

Rosenberg, A., (2000), "Reductionism in a Historical Science", *Philosophy of Science Asociation* 68: 135-146.

Simpson, G.G. (1940), "Mammal and Land Bridges", *Journal of the Washington Academy of Science* 30: 137-163.

Simpson, G.G. (1944), *Tempo and Mode in Evolution*, New York: Columbia University Press.

Simpson, G.G. (1949), *The Meaning of Evolution*, New Haven: Yale.

Simpson, G.G. (1953), *The Major Features of Evolution*, New York: Columbia University Press.

Simpson, G.G. (1961a), *Principles of Animal Taxonomy*, New York: Columbia University Press.

Sober, E. (1993), *Philosophy of Biology*, Boulder: Westview Press.

Spade, P.V. (1996), *Boethius Against Universals. The Arguments in the Second Commentary on Porphyr*, available in: <http://pvspade.com/Logic/docs/boethius.pdf>.

Suppe, F. (1998), "The Structure of a Scientific Paper", *Philosophy of Science* 65: 381-405.

Szathmáry, E. & J. Maynard-Smith (1995), "The Major Evolutionary Transitions", *Nature* 374: 227-232.

Valentine, J. (ed.) (1985), *Phanerozoic Diversity Patterns. Profiles in Macroevolution*, Princeton: Princeton University Press.

Wilkins, J.S. (1997), "A Taxonomy of Species Definitions (Or, Pophyry's Metatree)", New Zealand, *Works in progress*-version from 13.05.97, available in: <http://www.users.bigpond.com/ thewilkins /papers/metataxo.htm>.

Wilson, E.O. (1975), *Sociobiology. The New Synthesis*, Cambridge, MA: Harvard University Press.

Wilson, E.O. (1998), *Consilience: The Unity of Knowledge*, New York: Vintage Books.

When (We) Biologists Get Close to Philosophy in Order to Make Our Work Better. Epistemological Proposal in Parasitology and Reflections on Its Use[*]

Guillermo M. Denegri

Permanent Seminar on Bio-Philosophy, Faculty of Exact and Natural Sciences
Mar del Plata National University (UNMdP)/
National Scientific and Technical Research (CONICET)

1. A little bit of history...

I am about to finish my doctoral thesis in natural sciences and I came across a problem which I could only solve by appealing to the experimental method. However, I have chosen the most difficult way which has cause me to approach philosophy, only to discover a series of probably subtle (or irrelevant) inconsistencies for the professional parasitologists, but not for me. Was it possible that well-known parasitologists would gladly use ad-hoc hypotheses to solve problems of biological cycles of parasites, only to save their experiments? And even more: why did they propose those ad-hoc hypotheses knowing that they were violating basic rules of parasitology? Some of these, among other questions led me to a philosophy department to study

---

[*] The author thanks editors of this book for reading and providing useful changes.

and ask for help to solve problems in my routine as an experimental parasitologist.

As time passes by, I wonder whether my decision to study philosophy was the right one or if on the contrary it complicated my professional career and I had lost my chance of publishing more papers in specialized magazines which would have given me a better academic position. No doubt, I have not surrendered to temptation of giving up science; on the contrary, I have strengthened my vocation of biology, knowing that it is a paradigmatic discipline with interesting theoretical problems which needs a good dose of philosophical formation. As Bunge said (2003): "biology and philosophy, far from being different, partially overlap [...] and philosophy constitutes a bridge between sciences and humanities, and helps their development".

The main aim of this research work is to demonstrate how the meddling in philosophy contributed to make certain assumptions evident (which are sometimes not explained in the routine of a scientist) in my job as an experimental parasitologist. That is why I will demonstrate how an epistemological proposal helped me, and I will suggest how to make this proposal more useful from the point of view of a scientist of natural sciences which, eventually, I believe will make the work of a biologist and epistemologist better and richer (Denegri 2000, 2003a, Martínez, Denegri & La Rocca 2003). Among these elucidations I will demonstrate how the clearing-up of the biological cycle of an important parasite in veterinary medicine was delayed for three decades by the misuse of an ad-hoc hypothesis.

2. What parasitology is about and how a parasitologist works

Parasitology is a biological discipline whose main aim is the study of parasitical organisms. The biological association named parasitism may be defined as "an intimate relationship between two heterospecific organisms, during which the parasite, usually the smallest of the two species, metabolically depends on the host" (Cheng 1978). The relationship can be permanent (as in the case of taenias) or very short-lived (leeches, mosquitoes, ticks and others). Parasites metabolically depend on their hosts so that their relationship is obligatory Several types of parasitism can be distinguished: i) *facultative*: an organism whose life does not totally depend on parasitism, but that is able to adapt itself to it; ii) *compulsory:* when the organism depends completely on the host during a part or all of its vital cycle, and iii) *accidental:* when an organism starts an accidental relationship with its host which is not the specific one and survives.

According to its location it may be talked of in-parasites and out-parasites. The in-parasites live in the interior of the host's body, in places like the digestive tract, blood, tissues, etc. The out-parasites settle down in external surfaces of the host, or they are placed in superficial parts of their body. The host is usually the bigger of the two species which have the relationship. The hosts may be classified in: i) *definitive or final host:* which is the one where the parasite reaches its sexual maturity; ii) *intermediate host:* where the larval state is developed and iii) transport *or parataenic host:* it is the one which the parasite uses as a temporary shelter and it is a vehicle to reach the compulsory host, which is in general the definitive one.

As regards its organization we can speak of unicellular organisms (protozoa) and multi-cellular organisms (trematodes, cestodes, nematodes, acanthocephalans and arthropods).

Parasitology is a science which reaches its highest development from half of the XIX century and at present day it is an institutionalized discipline in all the university careers of biological formation (pharmacy, biochemistry, medicine, veterinary, agronomy and biology) and with a really good academic scientific prestige.

Although parasitologists work with a series of assumptions, which are not many times explained in his fieldwork and laboratory work, they are the ones which give unity and autonomy to the discipline and enable him to define it as a mature science. These assumptions are directly related with one of the most complete theories of biology such as the neo-Darwinian theory of evolution, which together with other minor theories, make Parasitology a consolidated discipline, and the one which offers (and in a way it does) a field of wide cross-curricular knowledge with the formation of inter-disciplinary and cross-disciplinary teams of work.

3. Starting point of the epistemological problem

I will demonstrate in an easy way a first approach to the empirical problem in parasitology which was the cause of my looking to epistemology and then the decision of appealing to an assumption which arose in the philosophy of the science as an answer to the questionings in the discipline. Moreover, after this introduction I will detail the elements which made it necessary the construction of a theoretical-methodological scheme in parasitology and its use and usefulness to the community of scientists of this science for their present and future theoretical and experimental work. I will also criticize the conventional use which is made of the different epistemological proposals which a priori seemed to be applicable to the analysis and development of the great divisions of the science (that may be physics, chemistry, biology, social sciences, etc.)

The cluster of compulsory parasites, which I worked with, belonged to the helminthes and within them to the flat worms popularly known as taenias or tapeworms (Denegri 1987, 2001, Denegri, Bernardina, Perez-Serrano & Rodriguez-Cabreiro 1988). They are the metazoan parasites with a greater degree of specialization. All the adult members are in-parasites of the digestive system and enclosed ducts of several vertebrates. During their vital cycle they need one, two or more intermediate hosts, in each of which they experience a stage of development.

The diversity of this cluster of parasites in nature is very large and there are more than 20 families which contain hundreds of genres and species. Among these families we find the Anoplocephalidae (the cause of this analysis) which feeds off a wide range of hosts that include the amphibians, reptiles, birds and mammals. They need the presence of an intermediate host and a definite host so as to complete their vital cycle. The intermediate host is small mites of the ground, of a free life and of a cosmopolitan distribution, named oribatids (Denegri 1993). The basic knowledge of the biology of these cestodes may be expressed in the following proposal: *the oribatid mites act as intermediate hosts of the cestodes of the Anoplocephalidae family*. The problematic situation emerging from this controversial biological cycle of *Thysanosoma actinides* (Denegri 1987), which is a parasite belonging to the biliary small ducts and the first centimeters of the duodenum of domestic and wild ruminants was the starting point for the incursion in epistemological waters. The experimental data brought forward by Allen (1973) and the following postulation of an *ad-hoc*[1] hypothesis to solve the biological cycle of this kind of parasites caused the questioning of the most elementary basis with which parasitologists work and arrange the explanatory and predictable schemes of the parasite-host relationship (Denegri 1991, 1996, Denegri & Cabaret 2002). Allen (1973) thought that this parasite needs two intermediate hosts and though he considered that this possibility was little probable, he explained that it should be taken into account and investigated.

The most striking feature is that taking Allen's work into account, all the texts about veterinary and biological parasitology refer (and even today) they are still mentioning a group of insects named psocopteros (evolutionary ancestors of the lice) as intermediate hosts of *T. actinioides*. Perhaps the crucial question would be to wonder why the biological cycle of *T. actinioides* has lost validity and has been almost forgotten for 30 years. No doubt, it is an interesting case for historians of science (and of parasitology) that it has been lately taken up again and that means a significant contribution from the theoretical-methodological point of view, since the empirical evidence suggests that what has been established about the biological cycle of this species is seriously questioned (Denegri, Elissondo & Dopczhiz 2002).

Allen's attempt to solve the problematic cycle of these species and the searching of a second intermediate host is a clear attempt to turn back to an ad-hoc hypothesis that is contradictory to the hard core which will defend this research work.

The searching of this ad-hoc hypothesis really contributed to expose the absence of a theoretical-methodological framework in parasitology. It was immediately attempted to develop an epistemological corpus which would have a specific application for the experimental and theoretical research in the discipline.

---

[1] According to Lakatos' terminology an *ad-hoc* hypothesis is the one which has an excess of empirical content but which has not been proved.

4. Methodology of research programmes applied to parasitology

According to what has been formerly stated in previous works (Denegri 1991, 1996, Denegri & Cichino 1997, Denegri & Cabaret 2002) the methodology of scientific research programmes – **SRP**– (Lakatos 1983) fits, prima facie, to postulate a parasitical theory applicable to in-parasites protozoa, metazoans (trematodes, cestodes, acanthocephalans and nematodes) and surface parasites (out-parasites).

I will support the methodology that the **SRP** is applicable to the characterization of theoretical developments with a strong empirical guarantee in present-day science. This methodology cannot only be used for the reconstruction of historical cases (Denegri 1997a) but it is also an important element for the help of the experimental researcher in his present-day job and in the design of future research work.

The novelty of the presentation lies in the fact that appealing to the methodology of Lakatos (with criticisms and modifications) a scientific research programme was built on parasitology with good heuristic power. The distance covered was contrary to what could be supposed: scientists did not reach the methodology trying to assemble a jigsaw puzzle together with the available hypotheses, and laws and theories in a certain field of scientific knowledge, but on the contrary, the hypotheses, laws and theories were set up in order to use Lakatos' proposition and develop the programme.[2]

The "hard core" of the programme of scientific investigation in parasitology is the following proposition: *the knowledge of the nourishing*

---

[2] In general when the epistemological literature and especially the use and /or application of different currents are analyzed (i.e. positivism, neo-positivism, hypothetical-deductive approach, etc.) to a certain scientific discipline, what can be observed is a reconstruction of these statements, trying to see if such a science fits to it or it does not. Moreover, the starting points are the scientific disciplines with more or less elaborated and contrasting theories and what is intended to be demonstrated is whether it follows the development put forward by the epistemological current at issue. That is why, in this work, we speak of an opposite way to the one to which we are used to as an epistemologist. From the scientific point of view the help of a theoretical-methodological proposal guided us to systematize and clarify judgments, hypotheses, laws and theories of a science like parasitology.

chains of the hosts (intermediate and definitive ones) allows us to explain and foretell the inside parasite fauna which they lodge.

This "hard core" is irrefutable by methodological decision of the community of parasitologists. Therefore it is necessary to elaborate a protective belt of auxiliary observable hypothesis to which we have to direct contrasts.

Taking as an example the cestodes parasites of the Anoplocephalidae family, this proposition allows us to explain the fact that sheep, bovines and equines are strictly herbivorous. These hosts only lodge parasites of the group of adult cestodes of the above- mentioned family and larval forms of cestodes whose definite hosts are carnivorous.

If the parasitical fauna of adults and larval cestodes which lodge these or other herbivorous is, in fact, not known, the hard core of the **SRP** (feeding behaviour) allows us to foretell which cestodes will be found.

It can be explained (and predicted) the biological cycle of a parasite by the simple fact of knowing the food chain of its host, and predict (and explain) the tropism of a host according to the knowledge of its parasitic fauna.

The **SRP** formed by the hard core plus the observation of auxiliary hypotheses and starting conditions explain and predict the registered and unknown parasitic fauna in future hosts to investigate.

The protective belt of auxiliary hypotheses is built (for the time being) on the following two hypotheses (Denegri 1991, 1996):
i. *hypothesis of biological cycles*;
ii. *hypothesis of the development of communities of parasites.*

The hypothesis i) of the biological cycles is built on the basis of the parasitological knowledge and it is expressed in the way of statements about the behaviour of each group of parasites: trematodes, cestodes, nematodes and acanthocephalans (Denegri & Cabaret 2002). The hypothesis ii) of the development of communities of parasites is based on different alternatives of evolution of communities of parasites expressed in four models drawn up from the parasitic ecology (Holmes & Price 1985, Price 1987, among others).

The auxiliary hypotheses of the protective belt are put forward at the same time as the programme moves on, but they cannot be completely built up *a priori*. These hypotheses can by themselves predict or explain phenomena and even join the new information to the body of **SRP**, which must necessarily interact between each other so as to get its aim of preserving and taking care of the hard core.

The starting conditions of the **SRP** are defined as necessary (physical) pre-conditions for the setting up of a parasite-host relationship and they are:

i. *existence of a potential parasite*: it is a parasitic species of another different host from the one which is considered for the analysis or a living species biologically free and able to capture a room in a living being.
ii. *existence of the potential host*: it is a species able to offer resources so that a parasite can fulfill completely or partially its biological cycle.
iii. *existence of a potential biotope*: where the members of the biological cycle of a parasite do not live together naturally but they have possibilities of survival in case of being introduced any of them, creating isolated phenomena which can generalize if the causes which produced them continue (Denegri, 1985).

A good **SRP** is the one whose main aim is to define qualitatively and quantitatively the potentiality of the phenomenon. The idea of potentiality of the parasitic phenomenon arises as a direct consequence of the structuring of this **SRP** in parasitology. The potentiality of the parasitic phenomenon has been defined as a real possibility that a parasitic organism has to conquer a room in a host (Denegri 2002).

The term potentiality denotes not only the possibility but also the probability that the phenomenon takes place. It can be explained and repeated that organisms that had a free life have progressively adapted to parasitism provided that the contact and frequency enough were produced for the relationship to maintain through time.

The term potentiality of the parasitical phenomenon, which infers from the programme of scientific investigation in parasitology, can help not only to elucidate multiple aspects that make to the study of the evolution of parasitism but explain other biological associations as well.

In a previous work (Denegri 1997b) it has been attempted to relate the concept of potentiality of the parasitic phenomenon with the theory of tendencies proposed by Popper with the purpose of having a wider conceptual framework when we refer to the term potentiality:

"*it is a process in display of possibilities to be carried out, open to new possibilities*".

Popper exposes a natural law which states that "all possibilities different from 0, even those to which we appoint to a lower propensity will end to be carried out as time passes by, as long as they have time enough to do it, that is to say, provided the conditions are repeated with enough frequency or remain constant through enough time" (Popper 1990).

Having said so and elaborated this **SRP** in parasitology we are able to express two empirically contrasting consequences:

1. Direct cycle parasites (monoxenos) are most frequently present in hosts of herbivorous diet. This last condition favors the larger density of these parasites and it is possible to foretell that there will be a greater association with others of the same cycle.
2. Indirect cycle parasites (heterogeneous nature) are present in hosts of carnivorous and omnivorous diet with a great variety of different species. On the other hand, the parasitic density in the host will be low.

If the **SRP** in parasitology is applied to the group of parasites that was originally the reason of the epistemological reflection it can be predicted:

i. The more herbivorous the diet is the more density of Anoplocephalid cestodes and the more diversity of species.
ii. The less herbivorous the diet is the less density of Anoplocephalid cestodes and the less diversity of species.
iii. Anoplocephalid cestodes are not found in carnivorous hosts, with the exception of a finding of a genre in pet dogs (a fact which is easily explained if one reads the research programme that we are presenting).
iv. Very low density Anoplocephalids and low diversity of species are occasionally found in omnivorous hosts.

5. How this proposal makes parasitology better

The purpose of a scientific research programme is to keep its productivity even when difficulties arise. Emerging difficulties are the ones which critically challenge the programme and give it the chance to improve, acting as potential opponents of its hard core. If the hard core is well built it will confirm itself appealing to its protective belt, which will strengthen each time more through new auxiliary hypotheses. It will happen if anomalies that challenge the programme are treated as soon as they appear, and they are not filed (Denegri 2003b). The treatment of these anomalies changes from this analysis and improves Lakatos' proposal in several aspects giving it greater effectiveness when it is applied to a certain field of scientific knowledge. Undoubtedly, we risk more if we consider and treat each anomaly as soon as it appears, but on the other hand, it provides the programme more security, increasing its explanatory and predictable power.

The peculiarity of the methodology of the **SRP** of making its predictability richer and the possibility of improving it from the facts that threaten the hard core, allows us to have a highly practical weapon. According to the above-mentioned the methodology of the programmes of scientific research must be taken into account by the scientists as a heuristic resource in their fieldwork and laboratory activities.

In order that a **SRP** in parasitology can be used by the parasitologists it must:

1. *provides the possibility of a parasite-host relationship:* which requires of a contact and can be in nature determined by biological factors of the species to relate. The preservation of the relationship and the terms in which it develops are determined in nature by particular environmental conditions and by observed variables (alternatives) and exemplified in the hypothesis about the development of communities of parasites. A necessary condition is the existence of tropic relationships (food chains) and a stable predator-prey chain, even in the cases of vectors.
2. *measure the possibility and the probability of the occurrence of a parasite-host relationship:*
    i. the possibility is given by the potentiality of the biotope and by biological characteristics of the species in question. Here the predation or tropic relationship does not play a role.
    ii. the probability of success depends on the frequency with which the challenges are produced. It is of ecological nature and therefore the tropism acts.

3. *to explain the causes and the colonization process to a host or to an environment:* the parasites can serve as indicators of recent colonization to certain hosts and this is explained according to a minimum or drastic variation in the diet, when changing from a biotope to another by modifications of the original biotope. According to this, the results of the colonization can be:
   i. parasites specialize or ii. they are an opportunist component in the new host. This distribution of the parasites can be explained by the historical co-evolution or by ecological factors. On the basis of the proposal put forward here, the ecological factors are the ones that can explain and predict the parasite-host relationship and only appeal to evolutionary explanations in second place. The specification phylogeny is a tautological expression: *"it is specific because it is specific"*. It shows an *"a posteriori"* historical process (Denegri 2003b).
4. *to explain and to predict changes in the parasite-host relationship:* a theory in parasitology whose hard core is based on the food chains of the hosts explains and predicts the potentiality of the phenomenon which must necessarily be mentioned referring to it by evolutionary and/ or genetic explanations.

On the basis of empirical works developed in parasitology and which act as contrasting to the proposal and defining a theoretical-methodological scheme whose starting budget (i.e. hard core) is the tropism of intermediate and definitive hosts plus the auxiliary observable hypotheses of the protective belt and the initial conditions, the following can be deduced:
i. Parasites serve as indicators of both present-day and past ecological interactions.[3]

---

[3] To contrast the hypothesis that that the parasites are useful indicators of ecological interactions in the past one can appeal to paleoparasitological data, among others, the study of coprolites, which allows us to analyze the parasitical fauna in hosts that lived thousand of years ago, and besides, become familiar with the characteristics and climate variations in the past, on the basis of the presence or absence of certain parasites when comparing them with the present day fauna. These paleoparasitological studies have fundamental importance for the analysis of the evolution of the genus *Homo* and other ancestor genres, to correlate them with the different surroundings and their variations in time. An additional datum that the analysis of the

ii. Parasites serve as indicators of recent settlements of hosts to new habitats.
iii. Parasites serve as evolutionary indicators (co-evolution) provided that it has been appealed to an ecological explanation, which implies tropic stability in time. Therefore, the stability or instability of the parasitical fauna of a host is essentially explained on the basis of trophic stability and instability and not to the philogenetics age of the host.

The interesting point of this **SRP** is that it not only has the advantage of offering a better theoretical-methodological structure of parasitology for its study but it also offers a framework to future experimental approaches, which will have the purpose, among other things, to test the hard core.[4]

6. Final considerations

A criticism to this proposal was that the parasitical problem, which is a special case of biological relationships, seemed too easy to apply Lakatos' scheme.[5] Nevertheless, I think that the suggested development (as an attempt to clear up the concept of parasitology) should serve as an example of how an "*a priori*" methodological proposal, that should be applied to the analysis and development of physics, chemistry, biology and social sciences, among others, has a definite

---

coprolites can provide is the composition of the diet of the hosts, and it could give a clue about their nutritional state.

[4] In general scientists work on a properly based programme which has demonstrated its productivity. It is worth having always in mind the risks that Popper has properly pointed out in his criticism to the standard science of Kuhn. Science should always be a critical activity, even though the experimental results continue corroborating the paradigm (or the hard core) in force. The fact of testing the hard core is not a contradictory expression to what we are referring to (and Lakatos himself proposed), this "so-called" testing means to show how powerful this hard core is and how it is coming out successful from the empirical confrontation and allows us to continue with the progress of the programme.

[5] This objection was raised with a constructive spirit by my close friend Dr. Carlos Castrodeza (Department of Philosophy, University Complutense, Madrid, Spain) who took the trouble to analyze my writings critically and point out possible weaknesses or potential objections to the proposition.

application in smaller plots of the scientific activity.[6] The fact of having used Imre Lakatos' methodology of the programmes of scientific research was not for reasons of epistemological fashions and as it has been intended to demonstrate in this work (and in former contributions) I dare to suggest little modifications to the original proposal of the Hungarian thinker. I am especially interested in parasitology as a science and my obsession is to learn it more and more to try to make it better from the theoretical and methodological point of view. Consequently, if in the future, this methodology stops being an interesting theoretical-methodological weapon to deal with the parasitical phenomenon, we would suggest (without any methodological prejudice) other epistemological alternatives.

A fact which makes the bibliography clear is that in several special fields, among which we may mention biology, there is a special predilection for Lakatos' proposal and it is shown in the literature of these last years (Michod 1981, Craw & Weston 1984, Denegri 1991, 1996, 1997a, Denegri & Cabaret 2002, Dressino, Denegri & Lamas 1998, Mangano & Buatois 2001, Fernandez & Gonzalez Sagrario 2003, La Rocca 2003, Zanetti & Blanco 2003, among others). The most important point of the reconstructions of some of the analyzed disciplines is that they show that the methodology of the programmes of scientific research surpasses the framework of a history (as it was also the main aim of Lakatos himself) to become a valuable guide in the present and future research (Denegri 2000).

References

Allen, R. (1973), "The Biology of *Thysanosoma actinioides* Diesing, 1834 (Cestoda Anoplocephalidae) Parasite of Domestic and Wild

---

[6] Even admitting that the parasitical phenomenon is an easy problem (something which I personally do not believe) to apply Lakatos' project to, what I have wanted to show is, on one hand, its usefulness for the theoretical and experimental parasitologists as a maker of a proposal for the work and the evaluation of research projects in the discipline; and on the other hand, to alert epistemologists on the use of methodological proposals in the most restricted areas of scientific activity. This last consideration may make scientists and epistemologists become friends in order to build fluent channels of communication and to be able to design joint work which will no doubt improve their activity (Denegri 2003a).

Ruminants", *Agriculture Experimental Station Bulletin*. N° 604. New Mexican State University, 68 pp.

Bunge, M. (2003), "Quien filosofa no está acabado", in Denegri, G. & G. Martínez (eds.), *Actualizaciones en Biofilosofía*, Mar del Plata: Editorial Martín, pp. 7-11.

Cheng, T. (1978), *Parasitología General*, Madrid: Editorial AC..

Craw, C. & P. Weston (1984), "Panbiogeography: a Progressive Research Program?", *Systematic Zoology* 33: 1-13.

Denegri, G. (1985), "Desarrollo experimental de *Bertiella mucronata* Meyner, 1895 (Cestoda: Anoplocephalidae) de origen humano", *Journal of Veterinary Medicine B* 32: 498-504.

Denegri, G. (1987), "Estudio sobre la biología de los cestodes anoplocefálidos que parasitan a rumiantes domésticos", *Tesis de Doctorado en Ciencias Naturales*, Facultad de Ciencias Naturales y Museo, Universidad Nacional de La Plata, N° 484, 56 pp.

Denegri, G. (1991), "Definición de un programa de investigación científica en parasitología: acerca de la biología de los cestodes de la familia Anoplocephalidae", *Tesis de Licenciatura en Filosofía*, Depto. de Filosofía, Universidad Nacional de La Plata, 64 pp.

Denegri, G. (1993), "Review of oribatid mites (Acarina) as Intermediate Hosts of tapeworms of the Anoplocephalidae", *Experimental & Applied Acarology* 17: 567-580.

Denegri, G. (1996), "La metodología de los programas de investigación científica aplicada a la parasitología", *Revista de la Asociación de Ciencias Naturales del Litoral* 27: 69-77.

Denegri, G. (1997a), "Contrastación de un Programa de Investigación Científica en Parasitología: reconstrucción de un caso histórico", *Natura Neotropicalis* 28: 65-70.

Denegri, G. (1997b), "La teoría de las propensiones de K. Popper y el concepto de potencialidad del fenómeno parasitario", *IX Congreso Nacional de Filosofía (AFRA), Libro de Resumen*, La Plata: AFRA, p. 12.

Denegri, G. (2000), "Hacia un entendimiento fructífero entre científicos y filósofos de la ciencia: un acuerdo civilizado sin exhabruptos", in Denegri, G. & G. Martínez (eds.), *Tópicos actuales en filosofía de la ciencia. Homenaje a Mario Bunge en su 80° aniversario*, Mar del Plata: Editorial Martín, pp. 79-96.

Denegri, G. (2001), *Cestodosis de herbívoros domésticos de la República Argentina de importancia en medicina veterinaria*, Mar del Plata: Editorial Martín.

Denegri, G. (2002), "El concepto de potencialidad del fenómeno parasitario y su aplicación al estudio de las relaciones parásito-hospedador: un análisis epistemológico", *Natura Neotropicalis* 33: 65-69.

Denegri, G. (2003a), "Breves reflexiones críticas sobre la utilidad de la epistemología para la tarea del científico profesional", *Nexos* 10 (17): 4-5.

Denegri, G. (2003b), "Programas de investigación científica e investigación experimental en biología", *Tesis de Doctorado en Filosofía*, Facultad de Humanidades y Ciencias de la Educación, Universidad Nacional de La Plata.

Denegri, G. & A. Cicchino (1997), "Un programa de investigación científico progresivo en parasitología a los ectoparásitos", *XIII Congreso Latinoamericano de Parasitología* (FLAP). Libro de Resúmenes, pp. 71-72.

Denegri, G. & J. Cabaret (2002), "La metodología de los programas de investigación científica como aporte epistemológico para la investigación experimental en parasitología", *Episteme* 14: 89-100.

Denegri, G., Elissondo, M. & M. Dopchiz (2002), "Oribatid mites as Intermediate Hosts of *Thysanosoma actinioides* (Cestoda:Anoplocephalidae): A Preliminary Study", *Veterinary Parasitology* 87: 267-271.

Denegri, G., Bernadina, W., Perez-Serrano, J. & F. Rodriguez-Caabeiro (1998), "Anoplocephalid cestodes of Veterinary and Medical Significance: A Review", *Folia Parasitologica* 45: 1-8.

Dressino, V., Denegri, G. & G. Lamas (1998), "¿Es posible una propuesta lakatosiana para el estudio del componente facial en mamíferos?", *Episteme* 3: 73-87.

Fernandez, M. & M. Gonzalez Sagrario (2003), "Lamarck: enemigo o precursor de Darwin?", in Denegri, G. & G. Martínez (eds),

Actualizaciones en Biofilosofía, Mar del Plata: Editorial Martín, pp. 113-127.

Holmes, P. & P. Price (1985), "Communities of Parasites", in Kikkawa, J. & D. Anderson (eds.), *Parasite Communitues: Patterns and Processes*, Oxford: Blackwell, pp. 187-213.

Lakatos, I. (1983), *La metodología de los programas de investigación científica*, Madrid: Alianza Editorial.

La Rocca, N. (2003), "La teoría de la evolución desde un enfoque lakatosiano", in Denegri, G. & G. Martínez (eds.), *Actualizaciones en Biofilosofía*, Mar del Plata: Editorial Martín, pp. 89-111.

Mangano, M. & L. Buatois (2001), "El programa de investigación Seilacheriano: la icnología desde la perspectiva de Imre Lakatos", *Asociación Paleontológica Argentina* 8: 177-186.

Martínez, G., Denegri, G. & N. La Rocca (2003), "Una instancia en la integración del pensamiento científico y el filosófico", in Denegri, G. & G. Martínez (eds.), *Actualizaciones en Biofilosofía*, Mar del Plata: Editorial Martín, pp. 15-45.

Michod, E. (1981), "Positive Heuristics in Evolutionary Biology", *British Journal for the Philosophy of Science* 32: 1-36.

Popper, K. (1970), "Normal Science and its Dangers", in Lakatos, I. & A. Musgrave (eds.), *Criticism and the Growth of Knowledge*, Cambridge: Cambridge University Press, pp. 51-58.

Popper, K. (1990), "A World of Propensities: Two New Views of Causality", in Popper, K., *A World of Propensities*, Bristol: Thoemmes, pp. 1-26.

Price, P. (1987), "Evolution in Parasite Communities", *International Journal for Parasitology* 17: 209-214.

Zanetti, M. & F. Blanco (2003), "La metodología de los programas de investigación científica de Lakatos y el desarrollo histórico de la biología molecular", in Denegri, G. & G. Martínez (eds.), *Actualizaciones en Biofilosofía*, Mar del Plata: Editorial Martín, pp. 227-241.

Montparnasse Station. Vindicating Jacob's Reductionism in Functional Biology

Gustavo Caponi

Philosophy Department, Federal University of Santa Catarina (UFSC)

1. Presentation

It has become common place to point out the contrasts between the experimental biology developed in laboratories, like that carried by Claude Bernard and André Lyoff, and the observational biology that stems from field work, like that carried out by Darwin and Niko Timbergen (see Allen 1979, 1994, Araujo 2001, Ricqlès 1996, Magnus 1997, 2000, Hagen 1999). By the same token, there has been a growing attention on a distinction between a reductionist approach of living phenomena and another approach that could be called, borrowing G. Gaylord Simpson's expression (1974, 1964, p. 42), a compositionistic one (*e.g.*, Jacob 1973, Pichot 1983, 1987, Morange 1994, 2002). I believe we should credit Ernst Mayr (1961, 1988) for having articulated very clearly both distinctions so as to place them into a more general contraposition between functional and evolutionary biology.

The former deals with proximal causes by means of predominantly experimental methods. Those causes have to do with individual organisms and explain how living phenomena hang together and combine themselves in the constitution of those structures. The latter, in turn, is intent on reconstructing remote causes, generally by means of comparative methods and historical inferences. Those causes operate on populations and explain why each of them evolves or has evolved the way they do or the way they actually did (Mayr 1980, 1985). It should be noticed that the point here is not to place in opposition of one another two alternative programs or paradigms; rather, what is worth highlighting is the distinction between two complementary ways of making inquires about living things. Such a distinction, I take it, is crucial for the consideration of a number of different problems in the philosophy of biology (Caponi 2001).

This is precisely the case of the controversy about the relationship between physics and biology. I believe this question should only be raised if it is specified which domain of biology we are referring to. Hence, taking this precaution into account and following the line of reasoning put forward by François Jacob in *The Logic of Life*, I shall limit myself in this article to examine and claim the legitimacy of a certain kind of priority from the molecular standpoint, not in the domain of the sciences of life in general, but rather in the specific case of functional biology. It is important to point out that the arguments we can devise for a certain kind of programmatic reductionism in connection with the latter case cannot be applied to the case of evolutionary biology. I also believe that, as far as evolutionary biology goes, we can and must hold a clear-cut anti-reductionism (Caponi 2001, 2002). However, I am convinced that, generally speaking, the vast majority of anti-reductionist arguments that can be found in current philosophy of biology refer to evolutionary biology. For this reason, they end up being either irrelevant or insufficient in order for us to understand the straightforward relationship that functional biology has always maintained with physics and chemistry. On the other hand, such a relationship has been stressed and highlighted by the developments of molecular biology that drive the progress of physiology and embryology and work as the ultimate court for any kind of dispute or controversy for all different fields of functional biology. This is the priority of the molecular viewpoint that I wish to address with the expression programmatic reductionism.

2. A few terminological explanations

The expression reductionism is definitely ambiguous: we can apply the term 'reduction' in many different ways. However, it is necessary to envisage a first approach to what we can usually understand by a reductionist stance or standpoint: it has to do with "the claim that objects or approaches of a certain kind can, at the end of the day, be defined or characterized by means of terms or components that correspond to another distinct approach" (Klimovsky 1994 p.275). Hence, in the field of philosophy of biology, and regarding the relationship between biology and physics, what is taken into consideration when we talk about reductionism is the possibility, or necessity, that biological phenomena or predicates can be defined, characterized or explained by our resorting to physical components, terms or theories.

However, as it has been acknowledged many times, this issue can be raised on distinct levels each of which suggests specific and somewhat autonomously treated questions. Actually, we can generally identify three levels of analysis each of which countenances either the vindication or the rejection of a certain kind of reductionism. Unfortunately, and considering the usual coincidence of the number three, there is no agreement about the terminology applied to characterize those levels, which can hide the fact that differences always emerge concerning the nature of the problem dealt with on one of the three levels.

First, we can refer to what is usually presented as the epistemological or theoretical plan or aspect of the reduction (Ayala 1983, p. 12, Dobzhansky *et al.* 1980, p. 485, Mayr 1988, p. 11). On this level, the one usually chosen by philosophers of science like Popper (1974, p. 187, 1983, p. 333), Nagel (1978, p. 312) and Hempel (1973, p. 152) in order to handle questions on reduction, what is at issue seems to be "whether experimental theories or laws devised in one realm of science can be considered as special cases of theories and laws devised in some other scientific realm". If this is so, we say that "the former branch of science has been reduced to the latter" (Ayala 1983, p. 12).

Nonetheless, apart from the first stage of the controversy about the possibility of reducing classic to molecular genetics, that standpoint has not called much of the attention of philosophers of biology (see Schaffner 1976, Hull 1974, Ruse 1979, Kitcher 1994, Gayon 1999). Nagel's conception of reduction, which was conceived of within the context of a reflection where physical theories have the last word, posed and trigged a controversy of local epistemology in connection with the relationship between two biological theories about heredity. Yet, such a controversy severely undermined, or at least called for a revision of, that idea of reduction as subsumption (see Schaffner 1993, Waters 1994, Callebaut 1995, Duchesneau 1997, Wimsatt 1998, Sarkar 1998).

It should be noted that the debate on the relationship between Mendelian and Molecular genetics, although it may be closely related to the aforementioned controversy, does not exhaust the debate about the relationship between biology and physics; actually, the debate in focus is not even a partial aspect of that relationship: in order for this latter to be possible, there should be presupposed that molecular biology itself is a chapter of physics or chemistry. As Rosenberg (1985, p. 43) claims, however, this would be a problematic simplification.

These points being noted, one thing is for sure: the revision and replacement of the problem of reduction that occurred at the heart of that controversy about genetics make us wonder if nobody would take very seriously nowadays the idea that the fundamental theories in biology can be reduced, in a very strictly Nagelian sense, to physical theories.

At the same time, while on the epistemological level, there is an almost general consensus among philosophers of science against Nagel's idea of reduction (Rosenberg 1994 p. 39, Sterelny & Griffiths 1999 p. 137); on the ontological level, we find a general reducctionist consensus. After the fall and subsequent departure from all forms of vitalism, there is no objection regarding ontological (Ayala 1983, p. 10) or constitutive (Mayr 1988, p.10) reductionism. Constitutive or ontological reductionism is that one considered by Dobzhansky (1983, p. 23) as reasonable, which is to say, the kind of reductionism that ends up acknowledging that all biological phenomenon or entity is, at the end of the day, a complex combination of physico-chemical phenomena and entities which, by way of this very combination, is subjected to that physical set of laws which govern its components

(Ayala 1983, p. 11, Mayr 1988, p. 11). To deny it is to embrace vitalism, or rather, it is to claim that living phenomena obey forces contrary or foreign to physical ones (Dobzhansky 1980, p. 486, Ayala 1983, p. 10, Mayr 1988, p. 10, Callebaut 1995, p. 37).

At this juncture, an explanation is in order: it does not seem correct to endow constitutive reductionism with the simple idea that organic phenomena are just limited by physical laws (Bauchau 1999, p. 237). The physical order is not only the condition of possibility and the realm within with life will build up an autonomous order. Life does not overcome physical order. It does not suppose any force that, by being alien to physical laws, would exercise a certain form of freedom in the constraints imposed by them. To deny vitalism, as Claude Bernard (1984[1865], pp. 120-122) claims, is to deprive living matter of all spontaneity, or to deny any capacity of change that does not require the intervention of a physical force (see Boutroux 1950[1893], p.72); in other words: to deny vitalism implies to adhere to the physicalist standpoint according to which there is no change, no difference in the world that does not demand a physical change or difference (Sober 1993a, p. 49). In the living realm, we could say that there is no efficient cause that is not a physical one.

That being the case, constitutive or ontological reductionism can be regarded as stating that all organic phenomenon physically recordable or observable, – i.e., all organic phenomenon that can interact with a physical instrument of observation and measurement – is thereby prone to a physico-chemical explanation. This is precisely what vitalism refused to accept: as per Stahl and Driesch, there were experimentally recordable phenomena that could not be explained in physico-chemical terms. Nevertheless, one thing is to say that all biological phenomenon that yield a physical sign or record in an instrument of observation can be described and explained in physical terms; and another thing is to state that all possible and relevant descriptions of a biological phenomenon can be translated into descriptions that can play the role of explananda of physico-chemical explanation.

At this point, we arrive at the debate on the third and most controversial form of reductionism: that one referred to by Mayr (1988, p. 11) as explanatory reductionism and by Ayala (1983, p. 11) as methodological reductionism. But, being that some commentators apply those two expressions in the same way as we have applied expressions like 'epistemological or theoretical reductionism' (see Klimovsky 1994, p. 283, and Dobzhansky *et al.* 1980, p. 485), we would rather call this last position programmatic reductionism. This is because we wish to highlight that this kind of reducctionism has to do with explanatory procedures and strategies (Dobzhansky *et al.* 1980, p. 485, Ayala 1983, p. 11); in other words: this viewpoint deals with the manner and the direction of investigation.

Nonetheless, beyond the terminological disputes, there can be a conceptual confusion that is not easily solved: it is not at all easy to explain what this new characterization of reductionism is about. Mayr (1988, p. 11), for example, identifies this thesis with the claim, in his false opinion, that all organic phenomenon can be explained in terms of the action and interactions of their components. But the point is not to go back to epistemological reductionism and claim that specifically biological theories that we currently use to explain organic phenomena are reducible or explainable by means of physical theories; rather, what is needed is to lay a wager on the possibility that, regardless of the theories we have taken for granted so far, organic phenomena can be analyzed and explained in strictly physical terms.

What is at issue is not the reduction of biological to physical theories, but rather, to replace the former with the latter. The difficulties of reductionism as it is conceived of by Nagel do not invalidate this strategy and, *despite* Mayr's approach, *constitutive reductionism* seems to back it up (Rosenberg 1985, p. 23), as we have just indicated. If organic phenomena are nothing but physical ones of great complexity, then there is no reason to give up Crick's purpose (1966, p. 10), namely, "to explain all biology in terms of physics and chemistry". However, to postulate this *physical explicability* of life is quite different from carrying it out. The postulation or the negation of a *physical explicability* of living phenomena touches on the need to talk about an *explicability-in-principle* and an *explicability-in-practice*.

Now, a word of caution is in order here: by *explicability-in-practice* we must not understand the mere effective and real capacity of providing a physical explanation to every biological phenomenon. If this were the case, the answer would be straightforward: there are uncountable biological phenomena we do not have a clue how to approach from a physical standpoint. It is also clear, though, that no one is interested in debating the question on such a level: the explicability of the reductionists will always be, in a trivial sense, an *explicability-in-principle*: a promise. A different problem arises when we wonder what physics this promise is based upon. From this point of view, which is the one Sober (1993, p. 25) most appropriately presents us with, *explicability-in-principle* means that an ideally complete physics would be equipped in such a way that it would account for all biological phenomenon; meanwhile, *explicability-in-practice* means that we can explain all biological phenomenon by means of the physics we already possess.

Proponents of the latter can ponder that our current incapacity in getting a grip on such explanations obeys our flawed analysis of biological phenomena and/or our ignorance on how to articulate the physical knowledge that does exist in an explanation of those phenomena. In turn, those who vindicate *explicability-in-principle* can argue that this incapacity stems from a constitutive or circumstantial limitation of physics. That being so, a proponent of *explicability-in-practice* can claim that this incapacity has to do with a limitation in biology and that it is therefore its subject matter and also its main concern to straighten out the situation. By the same token, those who resort to a limitation in physics to justify that incapacity of current biology will be acknowledging that, in order to rectify the situation, we should not look into the sciences of life themselves but rather expect a super-physics that is yet to come.

Keeping this in mind, the postulation of that *explicability-in-principle* yields less consequences to biology than to physics; hence, the debate on explicative reductionism should focus, at least initially, as it so happens, on the *explicability-in-practice*. Thus, the promise that underlies reductionist standpoints is the following: fundamental laws and forces being known, what should be studied is the complex web of initial conditions that make those forces and laws produce the phenomena of life. Science, borrowing Peter Medawar's celebrated expression (1969, p. 116), is the art of the soluble; to deny *explicability-in-practice* would place at least part of the reductionist endeavor beyond the limits of that solubility.

At the same time, to accept it seems to render such a kind of reductionism mandatory: if we start off by embracing the postulate that this reduction or explanation is possible, apart from actual difficulties, it is inevitable to commit ourselves to seeking it out. This is so not because a successful reduction is, as Popper holds (1984, p. 154), the most successful form whereby all scientific explanations can be conceived; rather, it is because to depart from the reductionist program would be equivalent to deny that current physics is enough to explain living phenomena. Better put, it would be like admitting that the current set of physical laws and forces is insufficient to explain life. If this were so, the old vitalism would be lurking around all over again and few of us would like to be considered as one of its followers. All of a sudden, then, the reductionist program can turn out to be as unobjectionable as constitutive reductionism.

However, if we remind ourselves of the last formulation we presented in connection with this kind of reductionism, we shall see that the relationship between the two theses is not as simple as one might expect. As we have pointed out earlier, one thing is to claim that "any biological phenomenon can be described as stemming from the interaction of physico-chemical processes" (Maturana & Varela 1994, p. 64); but to claim that all possible and relevant descriptions of a biological phenomenon can be translated into descriptions that can work as explananda of physico-chemical explanations is quite another.

Programmatic reductionism necessarily encompasses all biological phenomena just as long as they are describable or recordable in physical terms. If this were not so, we would be maintaining that there are physical phenomena to which there is no physical explanation at all. What is worth enquiring is whether the only relevant descriptions of a biological phenomenon are those liable to be converted into or replaced with descriptions that present them as mere physical events; or rather, whether there are descriptions that, by being biologically relevant, cannot be in principle translated into a physical language.

At this juncture, I take it, we should be careful to specify which one of those realms in the science of life, namely, either functional or evolutionary biology, we are referring to here. It is true from the start that the answer, affirmative or negative, that we could give to *functional biology* is significantly different from the one we could give to evolutionary biology. This can be easily noticed if we resort to the distinction between evolutionary and reductionist standpoints carried out by François Jacob (1973 [1970]) in his *Logic of Life*.

3. Jacob's reductionism

In this magnificent book, Jacob (1973, p. 14) turns on Mayr's distinction between functional and evolutionary biology and analyzes the two methodological stances that render possible, he believes, those two fundamental realms in contemporary biology. The first is an integrative and evolutionary stance that considers the organism as a member of a population and thereby countenances a biology the intent of which is to describe and explain the relationships that living things set up with their environment. The second, in turn, is the atomistic or reductionist stance that, although considering the organism as an individual whole, allows for a biology of proximal causes whose aim is to explain vital phenomena in terms of causal interaction of elements like organs, tissues, chemical reactions and molecular structures. As it is clear by now, in Mayr's language, the first one would be evolutionary biology and the second, functional biology.

Each one of those biologies, Jacob contends (1973, p. 16), "endeavors to set up an order in the living world". The first "has to do with the order by which living things are intertwined, the descent and the constitution of the species"; in other words, it is an inter-organic order. The second one, in turn, has to do with an intra-organic order that deals with structures, functions and activities by means of which the individual living thing is integrated and constituted. In this way, we can say that, if one "considers living things as elements of a vast system that encompasses all the earth"; the other one "is interested in the system that forms each living thing" (Jacob 1973, p. 16).

For this reason, whereas in this last case the biologist usually analyzes "a single individual, or a unique organ, a unique cell, a unique part of a cell" (Mayr 1998, p. 89); in the case of evolutionary or integrative biology, the organism has to be always considered in connection with its relationships with the environment and with other organisms (Jacob 1973, p. 14).

Keeping this in mind, whereas in the first realm of investigation we can still go on, to a certain extent, handling concepts and methods of natural history and with relative independence from physico-chemical knowledge (Jacob 1973, p. 200); in the second case, in turn, we find ourselves tackling a set of investigations that, by virtue of its own methodological agenda and in relation to the problems that were already been taken care of, gives rise to a discourse about living things that, because of its conceptual content and its experimental procedures, tends to zero in little by little the discourse of chemistry and physics. These areas would not reduce or absorb functional biology by broadening its field of application; rather, functional biology itself, by its own logic, would tend to actively identify its discourse with the ones found in those other sciences.

The main assumption of this strategy of investigation is that "there is no character of the organism that cannot, at the end of the day, be described in terms of molecules and their interactions" (Jacob 1973, p. 15). Following Jacob, we can consider this *programmatic reduction* as inherent to *functional biology*. However, beyond that assumption about the possibility of reduction, what truly defines that kind of reduction is the methodological requirement that, for all organic phenomenon, characteristic structure, we shall always look for a description and an explanation of physiological character such that they both can be reducible to physico-chemical descriptions and explanations. The successful achievement of such a reduction is based upon the aim that defines direction, agenda and criterion to evaluate the success of investigations developed in the context of *functional biology*.

Nevertheless, although this is a promise of molecular biology not very far from being fulfilled according to many (see Collins & Jegalian 2000), and without taking into account what can be said about *evolutionary biology*, it is quite possible that this *reductionist* image of *functional biology* is countered by means of three kinds of reasons that will be dealt with one by one. The first kind will be called *logical*, the second will be called *theoretical*, and the last kind will be called *historical*. It should be emphasized that those three lines of reasoning can be brought together on a single point: the idea according to which the main experimental task of *functional biology* is "to isolate the constituent parts of a living thing" so as to find out conditions that allow its investigation "in the test-tube" (Jacob 1973, p. 15).

Hence, Jacob says (Jacob 1973, p. 15), "by varying conditions, by repeating experiments and by zeroing in each parameter", the *functional biologist* would manage "to master the system and to eliminate its variables". The starting point of experimental work would be always, and without a doubt, the complexity of the individual living thing, but his aim would be precisely to decompose that complexity and to analyze its elements "through the ideal of purity and certainty that represent experiments in physics and chemistry" (Jacob 1973, p. 15). By following this analytical procedure, Mayr states (1998, p. 89), we can achieve in biology "the ideal of a purely physical, or chemical experiment".

Nevertheless, and this is the first line of objections that can be presented to this way of seeing functional biology, "not everything that occurs to an organism in the laboratory is a biological reality" (Merleau-Ponty 1953, p. 215); organic phenomena are worth considering just as long as they contribute, in normal state, or conspire, in a pathological state, to entail the constitution and preservation of the individual organism itself.

Without this interest, without this viewpoint that emphasizes and underlines the constitution and preservation of the individual organism as a converging point to all analyzed causal series, there is neither physiology nor *functional biology* in general. This discipline, I take it, always presupposed the idea of a privileged state and limits itself to determining how a certain organic phenomenon causally interferes in the production of such a state (Goldestein 1951, p. 340); that is precisely what Claude Bernard (1984 [1865], p. 137) pointed out so bluntly in the following passage of *Introduction a L'Étude de la Médecine experiméntale*:

> Neither the physiologist nor the physician can ever forget that the living thing constitutes an organism and individuality. The physicist and the chemist cannot place themselves outside that universe; they study bodies and phenomena one by one, on their own rights, without necessarily being bound to refer them to the totality of nature. The physiologist, on the contrary, locates himself on the outside of the living organism; he sees it as a whole and is concerned with its harmony in the same way as he strives to get into it so as to under-

> stand the mechanism of each of its parts. From this it follows that, whereas the physicist or the chemist can deprive the facts they observe of all ideas of final causes; the physiologist is led to concede a harmonic and pre-established purpose amongst organized bodies whose partial actions are all in synchrony with, and generated by each other. If this is so, it should be noted that we decompose living organism by breaking their parts away only with the aim of making experimental analysis easier, and not to conceive of them isolately. Actually, when we are intent on assigning value and true meaning to a physiological propriety, we must always go back to the whole and avoid drawing a definitive conclusion without consider its the relation with the whole.

The gathering together of living phenomena in functions, Bernard continues (1878, p. 34), "is the expression of that thought". In fact, the *function*, nothing but the privileged object of physiology (Coleman 1985, p. 241) and *functional biology* in general, is precisely "a series of acts or assembled phenomena that are harmonized with the purpose of bringing out a determinate result"; and although "the activities of a manifold of anatomic elements" are jointly acknowledged for the *function* to be carried out, such a *function* cannot be reduced to the "mere sum of the elementary activities of intertwined cells" (Bernard 1878, p. 370). Quite the contrary, in order to individuate the *function*, to render it possible to describe a set of organic activities as accomplishing a *function*, we ought to consider them as "harmonized, pieced together, so as to converge to a common result" (Bernard 1878, p. 370).

As Merleau-Ponty (1953, p. 215) would notice later on, "a total molecular analysis would dissolve the structure of the functions of the organism into the indivisible mass of trivial physico-chemical reactions". For this reason, "in order to make a living organism re-appear from them", we ought to re-assess those reactions by picking out "the standpoints from which certain sets receive an usual meaning and reappear, for example, as *assimilation* phenomena, like the components of a *function of reproduction*"; or definitely as moments or stages of any other function that our physiological analysis strives to set up.

Functional biology seems to assume, in fact, a point of view about organic phenomena that finds no counterpart in physics or chemistry. Likewise, for the reasons just shown, this approach in biology can be accurately called *functional standpoint*. Such a peculiarity, as Rosenberg

(1985, p. 39ss.) has argued, can also be singled out on the level of *molecular biology*. In front of a sequence of DNA, let us call it *gene*, the occurrence of which can be pinpointed in the genome of a certain species of bacteria, the molecular biologist will always ask for the *causal role*, or *function*, that the protein encoded by this gene plays in the constitution of such organism. This is exactly the sense Craig Venter presupposed when he stated that *forty five per cent* of individuated genes in the *Human Genome Project* still have an *unknown function* (see Gerhardt 2001).

There is a difference between asking for the mere *effect* of a phenomenon and asking for its *function*; such a difference lies in the fact that, in the latter case, we assume that the phenomenon can turn out to play a *causal role* (Cummins 1975, p. 745ss., Ponce 1987, p. 105ss., Caponi 2001, p. 39ss.) in the production of a certain state whose realization defines and orders our analysis. Although such an outcome may allow us to talk about a kind of *erothetic autonomy* of *functional biology* in relation to physics; I believe that outcome cannot be used to counter Jacob's *reductionism*.

It is true that, when he talks about analytical procedures of *functional biologist*, Jacob leaves out that *intra-organic teleology* alluded to by Bernard. But I believe that such carelessness is due to the simple fact that what is at issue is not the *way of enquire* in *functional biology*. Rather, what is to be dealt with here is the *ontological level* whereon the answers to the questions asked in that realm of biology should be looked for. The fact that the causal webs that constitute of individual organism are always rebuilt so as to bring out that result does not mean such a reconstruction should be prevented from being applied to the molecular field. After all, Jacob could concur with Maturana and Varela (1994, p. 18) about the idea that organisms are nothing but *molecular autopoiethic systems*, that is, physical systems capable of producing and preserving their own organization (Maturana & Varela 1994, p. 69). That would be a sound argument to keep up our endeavors to clarify the molecular mechanisms that materialize such *autopoiesis* (Maturana & Varela 1994, p. 68).

However, although we can say that the *functional standpoint* does not invalidate our carrying out the *reductionist program*, it is reasonable to pose a second objection against the latter that has to do with viability: that is the case of the *theoretical* objections propounded by Walter Elsasser (1998) and Jean Hamburger (1986), amongst others. According to this line of reasoning, the *reductionist program* could be viewed as posing, at least in areas like *developmental biology*, challenges that stand beyond our cognitive capacities. The point is not, of course, to doubt that organic phenomena are molecular ones. Rather, the point is to draw our attention to the computational complexity involved in any attempt to explain intricate physiological phenomena by means of molecular interactions. The variables to be considered and the possible interactions among them are so many and so complex that such a task seems to lie outside the mathematical limits of what is computable (Simon 1996, p. 172).

That is why, I take it, there is a common belief that some organic phenomena are best looked into through *classic* experimental strategies. Although they do not reach the molecular level, such strategies may countenance a much more significant control and knowledge of those phenomena.

After all, from Claude Bernard on, since the very birth of his realm, *functional biologist* thinks experimentally (see Bergson 1938, p. 230). Better put, he thinks of organic phenomena as long as he analyzes, controls and handles them experimentally. In his science, there is no room for causal links and concepts that do not conform to experimentation. This is also true for the *reductionist program*. Neither the vague certainty of *ontological reductionism*, nor the harsh demand of Sober's *explicability-in-principle* can stand on their own feet without having to express themselves operationally.

If we are indeed to account for biological phenomena in the scientific practice, to wit, as physical and chemical phenomena, this has to be done by means of experimental procedures; and what antireductionist arguments like *Elsasser's* hold is that, at least in the case of certain kind of phenomena, that is impossible *a priori*. If science is really the *art of the soluble*, the acknowledgement of such a limit may seem the ultimate condemnation for the *reductionist program*.

Now, this is precisely the weakness of those arguments: from the obvious fact that certain biological phenomena *are more complex than expected*; and from the difficulties stemmed from the successive attempts do analyze them from a purely molecular standpoint, there are voices who argue, from a mathematical perspective, that those difficulties are insurmountable and that any future attempt to overcome them are doomed to failure. The thing is, though, notwithstanding those difficulties, the reductionist program is alive and kicking: the laborious search and the hard charting of complex molecular thread wherewith complex organic phenomena are interwoven keep on bringing about results here and there. Although they are partial and almost feeble, such results are effective and cumulative (Rosenberg 2000).

The *reductionist program* seems to offer, at the end of the day, more tangible opportunities for the development of investigation than the mathematical limits alluded to by anti-reductionist theorists. Such heuristic fruitfulness, which is to say, the capacity to keep open the field of investigation by multiplying *jigsaw puzzles*, is the cornerstone of history of science (Chalmers 1979). Science is a practical activity and scientists always tend to follow ideas that can keep them busy (Gould 1995). This does not seem to be the case for Elsasser's *holistic* ideas.

It could be objected, however, that one thing is to say that the *molecular program* faces inexorable limits; but to say that the reduction of all organic phenomena to molecular ones is indeed the guideline for the development of *functional biology* is quite another. This is so because, without our resorting to the Elsasser's *non plus ultra*, we can confine ourselves to saying that perhaps Jacob is wrong to suppose that all *functional biology* zeroes in its own *molecularization*. Nobody would object to call this tendency a *hegemonic* one (Kitcher 1999; Lewontin 2000); but this does not mean to deny the existence of investigations that continue to produce new and significant knowledge, although they remain close to traditional physiology and embryology. Now we have arrived to the third and last set of objection that one can raise against Jacob's reductionism.

4. Montparnasse Station

The way to understand *functional biology* shown in *The Logic of Life* could be taken as a historical simplification: not every road in this realm drives us to Montparnasse Station. Not every line of investigation in *functional biology* seems to converge to *molecular biology* the foundations of which were set up by Lwoff, Jacob and Monod at Pasteur Institute, not very far from that station (Morange 2002).

As a matter of fact, although they are embedded with and instructed by knowledge about molecular components of cells, current experimental developments on cloning should be considered as the continuation of a program of *cellular engineering* the general traces of which were drawn independently of *molecular biology*. Experimental techniques developed therein allow for a control and manipulation of certain cellular processes whose molecular foundations are not thoroughly understood. This can be an indication of the feasibility and fruitfulness of research programs that are put to work on high level of organizations, like cells, tissues and organs. Those programs usually ignore the very molecular mechanisms upon which the phenomena that they study are based.

Nonetheless, in the face of the difficulties and failures that always come about in those realms of research; nobody would deny that the key to the hidden variables that debunk predictions and experiments could be found in the molecular structure of cells. One way or the other those research programs that, for pragmatic reasons, can develop with reasonable autonomy from molecular biology would go back to it in order to explain or to overcome its failures, as well as to base and explain its own successes. The point here is not to prophesy but to realize that every experiment in *functional biology*, since Claude Bernard, presupposes physico-chemical intervention on living things.

In those realms, biologists can reason from the point of view of either physiology or cells themselves; but they will *act* according to *physics*. That is why, at the end of the day, only by getting down to the molecular structure of living things can they pinpoint the ground of the explanation of the results to be found by means of their experimental interventions on the organic order. Although in practice and circumstantially there can be non-molecular programs in *functional biology*, its own nature and legitimacy depend upon our considering organisms as physical structures.

According to *functional biology*, *ontological reductionism* is more than a mere metaphysical conviction: its very possibility consists of the subjection of organic order to the physical one. Without that, the experimental project laid down by Claude Bernard (1984 [1865], pp. 144-

145) would be illegitimate. It should be noted that such a possibility also draws a limit and sets up an ultimate criterion of legitimacy to every explanation that can be devised in relation to an organic phenomena. An explanation like that, at least in principle, ought to be physically founded. *Functional biology* falls under the theoretical constrains of physics which, in turn, constitutes the highest tribunal we can resort to in the face of any unsolved conflict amongst concurrent biological explanations. Thus, in realms like physiology and embryology, the reduction of a biological statement or explanation to a physical one, although not always looked for systematically, turns into the ultimate criterion for the legitimacy of any particular result.

References

Adams, M.B. (ed.) (1994), *The Evolution of T. Dobzhansky*. Princeton: Princeton University Press.

Allen, G. (1979), "Naturalists and experimentalists: the genotype and the phenotype", *Studies in History of Biology* 3: 179-209.

Allen, G. (1994), "T. Dobzhanky, the Morgan Lab, and the breakdown of the naturalist/experimentalist dichotomy, 1927-1947", in Adams (1994), pp. 87-98.

Araujo, A. (2001), "O salto qualitativo em T. Dobzhansky: unindo as tradições naturalista e experimentalista", *História, Ciências, Saúde* 8 (3): 713-726.

Ayala, F. (1983), "Introduction", in Ayala & Dobzhansky (1983), pp. 9-29.

Ayala, F. & T. Dobzhansky (eds.) (1983), *Estudios sobre la filosofía de la biología*, Barcelona: Ariel.

Barreau, H. (ed.) (1983), *L'Explication dans les sciences de la vie*, Paris: CNRS.

Bauchau, V. (1999), "Emergence et réductionisme: du jeu de la vie aux sciences de la vie", in Feltz, B., Crommelinck, M. & Ph. Goujon (eds.) (1999), *Auto-organisation et Émergence dans les Sciences de la Vie*, Bruxelles: Ousisa, pp. 227-244.

Bergson, H. (1938), "La philosophie de Claude Bernard", in Bergson, H., *La Pensée et le Mouvant*, Paris: P.U.F., pp. 229-238.

Bernard, C. (1984 [1865]), *Introduction a l'étude de la médecine experimentale*, Paris: Flammarion.

Bernard, C. (1878), *Leçons sur les Phénomènes de la vie communs aux animaux et aux végétaux*, Paris: Baillière et Fils.

Boutroux, E. (1950 [1893]), *L'Idée de Loi Naturelle dans la Science et la Philosophie Contemporaines*, Paris: Vrin.

Callebaut, W. (1995), "Réduction et explication mécaniste en biologie", *Revue philosophique de Louvain* 93 (1-2): 33-55.

Caponi, G. (2002), "Sobreviniencia de propiedades e identificación funcional de propiedades en biología", in Cupani, A. & C. Mortari (eds.), *Linguagem e Filosofia*, Florianópolis: NEL-UFSC, pp. 191-201.

Caponi, G. (2001), "Biología evolutiva *vs.* Biología funcional", *Episteme* 12: 23-46.

Chalmers, A. (1979), "Towards an Objectivist Account of Theory Change", *British Journal for the Philosophy of Science* 24: 227-233.

Coleman, W. (1985), *La biología en el siglo XIX: problemas de forma, función y transformación*, México: Fondo de Cultura Económica.

Collins, F. & K. Jegalian (2000), "Le code de la vie déchiffré", *Pour la Science* 267: 46-51.

Creath, R. & J. Maienschein (eds.) (2000), *Biology & Epistemology*, Cambridge: Cambridge University Press.

Crick, F. (1966), *Of Molecules and Men*, Seattle: University of Washington Press.

Cummins, R. (1975), "Functional Analysis", *Journal of Philosophy* 72: 741-765.

Dobzhansky, T. (1983), "Comentarios preliminares", in Ayala & Dobzhansky (1983), pp. 23-24.

Dobzhansky, T., Ayala, F., Stebbins, L. & J. Valentine (1980), *Evolución*, Barcelona: Omega.

Duchesneau, F. (1997), *Philosophie de la Bioloogie*, Paris: P.U.F.

Elsasser, W. (1998), *Reflections on a Theory of Organisms: Holism in Biology*, Baltimore: The John Hopkins University Press.

Feltz, B. (1995), "Le réductionisme en biologie: approches historique et épistémologique", *Revue philosophique de Louvain* 93 (1-2): 9-32.

Gayon, J. (1999), "La génétique mendélienne a-t-elle été réduite par la biologie moléculaire?", *Biofutur* 189: 43-44.

Gerhardt, I. (2001), "Número baixo de genes é surpresa", *Folha de São Paulo*, April 12.

Goldstein, K. (1951), *La Structure de L'organisme*, Paris: Gallimard.

Goodfield, J. (1983), "Estrategias cambiantes: comparación de actitudes reduccionistas en la investigación médica y biológica en los siglos XIX y XX", in Ayala & Dobzhansky (1983), pp. 98-127.

Jacob, F. (1973), *La lógica de lo viviente*, Barcelona: Laia.

Hagen, J. (1999), "Naturalists, Molecular Biologists, and the Challenges of Molecular Biology", *Journal of the History of Biology* 32 (2): 321-341.

Hamburger, J. (1986), *Los límites del conocimiento*, México: Fondo de Cultura Económica.

Hempel, C. (1973), *Filosofía de la Ciencia Natural*, Madrid: Alianza.

Hull, D. (1974), *Philosophy of Biological Science*, New Jersey: Prentice Hall.

Kitcher, P. (1994), "1953 and All That: A Tale of Two Sciences", in Sober (1994), pp. 379-400.

Kitcher, P. (1999), "The Hegemony of Molecular Biology", *Biology & Philosophy* 14: 195-210.

Klimovsky, G. (1994), *Las desventuras del conocimiento científico*, Buenos Aires: A-Z.

Lewontin, R. (2000), *The Triple Helix*, Cambridge, MA: Harvard University Press.

Magnus, D. (2000), "Down the Primrose Path: Competing Epistemologies in Early XX Century Biology", in Creath & Mainschein (2000), pp. 91-121.

Magnus, D. (1997), "Heuristics and Biases in Evolutionary Biology", *Biology & Philosophy* 12 (1): 21-38.

Martínez, S. & A. Barahona (eds.) (1998), *Historia y explicación en biología*, México: Fondo de Cultura Económica.

Maturana, H. & F. Varela (1994), *De máquinas y seres vivos*, Santiago de Chile: Universitaria.

Mayr, E. (1961), "Cause and Effect in Biology", *Science* 134: 1501-1506.

Mayr, E. (1980), "Some Thoughts on the History of the Evolutionary Synthesis", in Mayr, E. & W. Provine (eds.), *The Evolutionary Synthesis: Perspectives on the Unification of Biology*, Cambridge, MA: Harvard University Press, 1998, pp. 1-50.

Mayr, E. (1985), "How Biology Differs from the Physical Sciences", in Depew, D.J. & B.H. Weber (eds.), *Evolution at a Crossroads*, Cambridge, MA: MIT Press, 1985, pp.43-63.

Mayr, E. (1988), *Toward a New Philosophy of Biology*, Cambridge, MA: Harvard University Press.

Mayr, E. (1998), *O Desenvolvimento do Pensamento Biológico*, Brasilia: Universidade de Brasilia.

Medawar, P. (1969), *El arte de lo soluble*, Caracas: Monte Ávila.

Merleau-Ponty, M. (1953), *La estructura del comportamiento*, Buenos Aires: Hachette.

Morange, M. (2002), *Monod, Jacob, Lwoff: les mousquetaires de la nouvelle biologie*, Paris: Pour la Science.

Morange, M. (1994), *Histoire de la biologie moléculaire*, Paris: La Decouverte.

Nagel, E. (1978), *La estructura de la ciencia*, Buenos Aires: Paidós.

Pichot, A. (1987), "The Strange Object of Biology", *Fundamenta Scientiae* 8: 9-30.

Pichot, A. (1983), "Explication biochimique et explication biologique", in Barraeau (1983), pp. 69-106.

Ponce, M. (1987), *La explicación teleológica*, México: UNAM.

Popper, K. (1974), *Conocimiento objetivo*, Madrid: Tecnos.

Popper, K. (1984), *El universo abierto*, Madrid: Tecnos.

Ricqlles, A. (1996), *Leçon inaugurale de la Chaire de Biologie Historique et Évolutionnisme*, Paris: Collège de France.

Roger, J. (1983), "Biologie du fonctionnement et biologie de l'evolution", in Barreau (1983), pp. 135-160.

Rosenberg, A. (1985), *The Structure of Biological Science*, Cambridge: Cambridge University Press.

Ruse, M. (1979), *La filosofía de la biología*, Madrid: Alianza.

Sarkar, S. (1998), *Genetics and Reduction*, Cambridge: Cambridge University Press.

Schaffner, K. (1976), "The Watson-Crick Model and Reductionism", in Grene, M. & E. Mendelsohn (eds.), *Topics in Philosophy of Biology*, Dordrecht: Reidel, 1976, pp. 101-127.

Schaffner, K. (1993), *Discovery and Explanation in Biology and Medicine*, Chicago: The University of Chicago Press.

Simon, H. (1996), *The Sciences of the Artificial*, Cambridge, MA: The MIT Press.

Simpson, G. (1974 [1964]), *La biología y el hombre*, Buenos Aires: Pleamar.

Sober, E. (1993), *The Philosophy of Biology*, Oxford: Oxford University Press.

Sober, E. (ed.) (1994), *Conceptual Issues in Evolutionary Biology*, Cambridge, MA: MIT Press.

Suárez, E. & S. Martínez (1998), "El problema del reduccionismo en biología: tendencias y debates actuales", in Martínez & Barahona (1998), pp. 337-370.

Waters, K. (1994), "Why the Antireductionist Consensus Won't Survive the Case of Classical Mendelian Genetics", in Sober (1994), pp. 401-418.

Wimsatt, W. (1998), "La emergencia como no-agregatividad y los sesgos reduccionistas", in Martínez & Barahona (1998), pp. 385-418.

About the Risks of a New Eugenics

Héctor Palma

National University of General San Martín (UNGSM)

Eduardo Wolowelsky

University of Buenos Aires (UBA)

1. About the so-called "New Liberal Eugenics"

The growing development of technologies associated with human reproduction has conveyed a new impulse to the debate (it would probably be more appropriate to speak about multiple debates or multiple levels of analysis) around the legitimacy of modeling the genetic configuration of human beings. In the past years, such intervention in the characteristics of future persons was presented in terms of molecular biology and genetic engineering and the current disputes arise as a dilemma between what we are able to do – or presumably are able to do – and what we should do. Considering the present state of our technological and scientific development, we can very likely do much less than we believe we can or mass media divulge, which explains, at least partially, the effectiveness of certain ethical decisions. In any case, it seems possible now, and to a greater extent in the future, to manipulate our future offspring significantly – with some evolutionary cost that is difficult to assess. In this context and against the background of the dramatic facts that occurred in the 20th century, the ghost of eugenics reappears, now under the name of "new" or "liberal" eugenics. The debate on new eugenics is always engaged in on the basis of certain consensus about the totally abusive and negative character of the eugenics already known. In this sense, those who support the new reproductive technologies try to point out the differences while those who condemn them warn the public against the similarities and potential risks.

In discussing actual eugenics, a somewhat broad definition is considered: eugenics is "any intervention, individual or collective, with a view to the modification of the genetic characteristics of the offspring, independently of the therapeutic or social goals pursued" (Soutullo 2001). The intervention in the characteristics of the future offspring is an additional step taken in connection with infertility problems, and the eugenic feature arises from addressing the issue of the "quality" of reproduction. Eugenics is "liberal" when the decisions on the future offspring are made by parents and not on state or institutional levels.

Our purpose is to contribute to clarify the premises of the debate, reflecting on whether the reappearance of eugenic programs, called liberal eugenics, may be feared in any sense. For this purpose, we will attempt to show the differences between some of the main features present in the eugenic process towards the end of the 19th century and in the beginning of the 20th century and the possibilities, nowadays, of manipulating the future offspring.

One of the most important issues in this debate that involves the marriage between genetic intervention technologies and reproductive medicine is the pre-implantation genetic diagnosis (PGD).[1] The PGD (see Testart & Godin 2001) enables the screening of chromosome conditions and certain genetic characteristics in embryos obtained by *in vitro* fertilization. As this screening is carried out before the transference of the embryo into the uterus, it gives the chance to select the embryos that will be used, which permits a selection process that may result in a eugenic interference. In the first place, we have an *in vitro* fertilization procedure. After the fertilization, the human zygote di-

---

[1] Other technologies, specifically those that may appear in the future, will not modify substantially the terms of the debate that we propose in this paper.

vides itself every 24 hours approximately, so that, 3 days after obtaining the ovules, the embryos have 8 cells or blastomeres approximately. At this point, 1 or 2 blastomeres are extracted by means of a micropipette (35 microns), without negatively affecting the subsequent embryo development. This technique, known as embryo biopsy, makes it possible to obtain a small sample of every embryo and its later analysis through cytogenetic and cell biology highly specialized techniques. By means of this procedure it is possible to analyze chromosomal numeric anomalies, such as the presence of three chromosomes 21, responsible for Down's syndrome. It also enables the study of structural chromosomal anomalies, mainly translocations. It is also possible to identify the sexual X and Y chromosomes and thus to determine the gender of the embryo, which is important in connection with sex-linked disorders, due to the fact that the relevant alleles are found in the X chromosome. Further, it is possible to amplify the specific sequence of DNA in which the presence of mutations can trigger a disease of genetic origin. Several thousands of genetic diseases have been described, such as cystic fibrosis, myotonic dystrophy, Tay-Sachs disease, beta-thalassemia, sickle-cell anemia, Huntington's disease, etc. These techniques of chromosome or genetic analysis make it possible to complete a diagnosis very rapidly, from 3 to 48 hours according to the case, which is compatible with the maximum time period of the *in vitro* embryo development. Accordingly, the embryos can be maintained in a growth medium before the results are obtained and consequently the embryos which will be transferred into the uterus can be selected.

PGD techniques offer the certainty to detect and eliminate serious diseases but, at the same time, it is possible to think that it is the threshold of a new selective eugenics (see Habermas 2001), which conjures the ghost of the horrors of the nazi regime, among others. It is appropriate to mention now the traditional distinction between negative and positive eugenics. The former is governed by a therapeutic or healing logic and can be defined as the attempt to eliminate, or reduce the frequency of appearance of, alleles that are considered prejudicial or dangerous for human beings or, at least, some particular population. On the other hand, positive eugenics is directed to promote the reproduction of certain individuals with some characteristics considered desirable. Both negative and positive eugenics have been associated with the implementation of practices and politics which had the objective of modifying the population composition, that is to say to influence the evolutionary process. Initially, the difference is definite: at the farthest extreme of negative eugenics there is the possibility of reducing or eliminating very serious diseases that cause great suffering and severely limit the chance of living an autonomous or even partially autonomous life. At the farthest opposite extreme, positive eugenics displays the superior race delirium. Nonetheless, this definite difference between both types of eugenics does not bear much relevance and melts into the so called new or liberal eugenics, which is based on the possibility of the parents, through the technology available, to intervene (modify, eliminate) in one or more characteristics of their children. Liberal eugenics, which is based on the decision of individuals, seems not to place any obstacle in the transition from a negative selection of embryos that are certain to present serious hereditary diseases to a selection of embryos according to desirable characteristics not related to any pathology.[2] There are many reasons for this, but the main reason is the individual decision-making and there is nothing to prevent what is abhorrent today from becoming usual tomorrow. It should not be overlooked that the concept of "disease" is contextual and highly variable. In fact, allegations in support of eugenics are nearly always based on the elimination of what is "inferior" and "pathologic". In the same way, we should bear in mind that the genetic system cannot always be considered as a sum of characteristics which are independent from each other and from the environment. Regarding this, cystic anemia is a paradigmatic example.

Let us state our problem again: is this "eugenic" position based on the parents' decision really "eugenic" in the same sense as the development of eugenics at the end of the 19th century and during the first half of the 20th century?[3] In order to answer this question, it becomes necessary to review some characteristics of eugenics.

---

[2] It is on this impossibility to make a distinction, that is to say the possible excesses, that Habermas (2001) bases his arguments against a possible 'liberal eugenics'.

[3] We might be providing additional arguments for those who support eugenics, in the sense that new eugenics is based on phenomena that differ from those already

2. The eugenic movement

The creation of eugenics is usually attributed to Sir Francis Galton (1822-1911), who defined it as the science that deals with all influences that improve the innate qualities – raw material – of a race and those other qualities that can be developed up to their highest point (Galton 1883, pp. 24-25). It is clearly seen that eugenic literature shows a continual confluence of two elements: on the one hand the traditional dreams of the artificial improvement of the human race, imbibed by the ideals of progress that were already and securely established in the Western World towards the end of the 19th; on the other hand, the belief, influenced by the theory of evolution, that modern conditions (the science of medicine, assistance programs, the "comfort" of modern life, etc.) contribute to prevent the effects of natural selection through the death of the less fit individuals, which could result in the decadence of the species (or race, according to the case).

Standard historiography considers that eugenics is a pseudo-science,[4] however, how should we designate a movement of ideas that has developed throughout the world and captured the attention of a vast majority of the scientific community, including biologists, medical doctors, geneticists, demographists, jurists, psychiatrists and psychologists among others, many of them Nobel Prize winners, that has led to the creation of national associations in most countries and international federations, and to the meeting of many conferences in which renowned scientists took part, that has established itself in universities; that has caused a huge number of writings to be published in specialized journals and magazines over at least 50 years, that, in spite of being a not totally uniform movement, has established the agenda on issues such as health, public hygiene and, many times, state policies? On the other hand, it does not seem accurate to consider that eugenics is a scientific theory in any of the senses in which this concept has been used in epistemological disputes. Besides, an examination of major eugenicists' thought shows that many of their assertions

---

known. It is a contribution, in any case totally unintentional, and it suffices to point out that there are strong arguments against a new eugenics.

[4] With differences some times significant, this is the view shared by Chorover (1979), Gould (1981), Kevles (1995), Hobsbawn (1987), Randall (1981), Bernal (1954). For a discussion on this point, see Palma (2002).

on heredity were *ad hoc* assumptions. Nevertheless, some Mendelian geneticists try to establish an empirical base for their assumptions by means of cytological studies, experimental crossings with mammals that exhibited some similar characteristics to those studied in humans.

Thus, it seems more appropriate to characterize eugenics as a set of actions of a technocratic and authoritarian nature (see Thuiller 1988) associated with the scientific knowledge available, many times implemented through active public policies and intended to favor the reproduction of certain individuals or human groups that are considered as superior and to inhibit the reproduction of other groups or individuals considered inferior or undesirable, with a view to the improvement-progress of humanity or those human groups. This broad characterization may be applied to every actual form adopted by eugenics in the first half of the 20th century in nearly all of the Western World in which eugenics was a comprehensive and generalized plan for the implementation of public policies which was carried out coercively and, essentially, eugenics did not consist of voluntary individual actions.

It is correct to attribute the fatherhood of eugenics to Galton on the condition that it is taken into account that his proposal came to materialize the development of widespread beliefs and aspirations in the end of the 19th century. It would be better to say that Galton's proposal fell in fertile soil because eugenics was based on assumptions shared by vast sectors of scientists and thinkers. Far from being an isolated or marginal belief, it rapidly gained high standing, and became the ultimate "scientific" rationale for health policy measures, but also it generated and strengthened usual believes and prejudices. Its influence and strength was felt to a lesser or greater extent throughout Europe, America and even some Asian countries. Racialism[5] and the concept of the degeneration of lower classes, widespread ideologies, caused the problematics prevailing in expanding cities to be interpreted as a "degeneration" process in progress. The idea of racial im-

---

[5] The concept of 'racialism' (Todorov 1989) differs from 'racism' in that the latter refers to a more or less spontaneous and generalized conduct of rejection and fear of persons that are different or foreigners; such conduct stems from prejudices present in common sense, whereas the former seeks the support of 'scientific' theories.

provement associated with health was based on the new genetic theories of the first years of the 20th century and the rise or fall of nations would depend on the application of them. The eugenic theory enjoyed such scientific authority and political influence that it was finally institutionalized by means of the creation of solid scientific societies throughout the Western World. It can be asserted that, even though the theoretical developments, scope and speculations of eugenics have been to a certain extent heterogeneous, the principal ideas of eugenics have cut across all aspects of the scientific, social and cultural life from the end of the 19th century through the first half of the 20th century, comprising points of view coming from biology, sociology, medicine, health policies, educational technologies, immigration policies, demography, psychiatry, law sciences and criminology. A large part of the developments in these areas in the first half of the 20th century take on a meaning within the frame of eugenic ideals (Massin 1991, Thuillier 1988, Kevles 1995).

Although eugenics is usually associated with nazi Germany, the eugenic movement had not only arisen previously but also had rapidly and widely extended over nearly all of the Western World. All over Europe eugenic institutions proliferated (Massin 1991, Thuillier 1988, Kevles 1995, Stepan 1991): the Eugenic Committee of The Hague was created in 1912 and transformed into the Eugenic Society eight years later; also in 1912, the Italian Society of Genetics and Eugenics was created, and in 1913, the Eugenic Society of France, the Eugenic Division of the International Institute of Anthropology of Paris, the Federation of Rumanian Society of Eugenics and the Catalan Society of Eugenics were born. In 1934, an International Conference on Eugenics was held in Zurich. Sweden and Russia had also their eugenic societies and, according to *La Semana Médica* (Kehl 1926, p. 480), the Hindu Eugenic Society was created in India.

Also Latin America echoed the eugenic ideals and proposals. In 1917, furthered by Renato Kehl, the Eugenic Society of San Paulo was created, the first one in Brazil and Latin America. In 1931 the Mexican Society of Eugenics was created in Mexico. In Cuba, Domingo Ramos, who coined the Spanish term *"hominicultura"* ("hominine culture"), disseminated and practised eugenics. In 1939 in Peru, the First Peruvian Conference of Eugenics was held. Argentina was a leader of Latin American eugenics, together with Brazil. Argentina had already sent delegates to the II International Conference on Eugenics held in New York in 1916. However, from the point of view of its institutionalization, the first important event goes back to 1918 when Dr. Víctor Delfino created the Argentine Society of Eugenics, which had a short existence. Afterwards, in 1921 Dr. Alfredo Verano created the Argentine League of Social Prophylaxis, and finally in 1932 the Argentine Association of Biotypology, Eugenics and Social Medicine, which published its "Annals" for many years. The Association had its own hospital and training institute. A clear understanding of the broad and extended character of eugenics, which comprised a huge number of issues, can be grasped if we take into consideration the thematic areas covered in the "Annals". It can be read on the back of every issue that the publication included works on: "constitutional medicine, endocrinology, biotypology, eugenics, social medicine, dietetics and nutrition, hygiene, health engineering, psychology, pedagogic education, physical education, criminology, social doctrine and legislation." All these institutions were in turn affiliated to the International Latin Federation of Eugenics Societies with offices in Paris, which favored the organization of the first Latin Conference of Eugenics that was held in August 1937. Three Conferences on Eugenics and Hominine Culture were held in the American Continent, the latest of them in Bogotá in 1938 (see Mac-Lean y Estenós 1952, and Plotkin 1996).

All these associations are a consequence of the strengthening of the eugenic ideals – on many occasions after years of efforts. By 1930, the Canadian state of Alberta (the destination of "irredeemable alcoholics") Denmark and Finland had passed sterilization laws following the American example. In 1934, a statute was passed in Sweden, at the proposal of the social democrats, that made mandatory the sterilization of any persons that were incapable of educating their children. In 1941, the sterilization laws included the asocial and undesirable persons, from mothers of many children to young persons with conduct problems who were committed to correctional institutes. One of the countries that was a leader of the eugenic movement was the United States. In 1910 the Office of Eugenic Reports was created and it gathered scientists from diverse fields so that they would study, report and recommend measures of a public character in connection with any matter related to their common objectives. The Unites States became the first nation in modern times to promulgate and apply laws that promoted eugenic sterilization in the name of the "purity of the white race". In Indiana, because of the significant immigration of black people, a statute was passed in 1907, which restricted immigration and promoted the sterilization of "social misfits". In the following years, seven more states of the United States promulgated legislation of this type and in 1915 twelve states had legislated in the same sense. Some sterilization laws, like Virginia's, were in force from 1924 to 1972, which permitted the performance of 7,500 operations on white men and women and on children with misconduct problems and on the grounds of alleged feeblemindedness, antisocial conduct or imbecility, according to the traits shown by IQ tests.[6]

2.1. Eugenic practices

There exists a battery of practices and typical social technologies associated with eugenics: mandatory prenuptial medical certificates, birth control, sterilization of certain groups ("feebleminded persons" and/or criminals), eugenic abortion, immigration restrictions. The extent of and strictness in the application of these measures have varied in different countries and times, and in some cases a part of the eugenicists' demands have not been implemented effectively. Usually, implementation has not gone beyond the mandatory prenuptial certificate and health control, while on some occasions it has reached unprecedented levels of brutality, like in nazi Germany and other European countries or some states of the United States (see Chorover 1979 and Gould 1981).

Probably, one of the more extended and comprehensive practices stemmed from eugenic advocacy is the mandatory prenuptial medical certificate, adopted gradually in practically all countries of Europe and America. In most of them the certificate was mandatory and the problematics of the control of the offspring through this mechanism established itself strongly and extensively. Between 1910 and 1935, practically all countries legislated on the matter. Nevertheless, some

---

[6] About the relation between IQ tests and eugenics, see Chorover (1979) and especially Gould (1981) and Taylor (1980).

countries chose a policy of dissemination and propaganda, and the certificate was optional. The Eugenic Society of London, for example, objected to the medical certificate being mandatory in Great Britain, and something similar occurred in the Netherlands, in which the Eugenic Committee of The Hague financially supported prenuptial dispensaries and policlinics in addition to conducting extensive educational campaigns, and a similar course of action was adopted by Italy. The requirement of obtaining the certificate was based on the concept that, in most cases, marriages were "uneugenic" and could cause the "degeneration of the race". The extent of the prohibition to get married was variable, but it was generally intended for alienated and mentally retarded individuals; persons suffering from syphilis, tuberculosis and alcoholism; but sometimes it was requested that the prohibition would apply to "sexual inverts".

Another flag that was waved by eugenics was birth control or, as was usually designated, the "scientific" control of conception. In any case, it was a "differential control of conception", since it was not merely directed to maintain birth rates at certain levels but to prevent or reduce the reproduction of certain groups. The implementation of contraceptive devices, quite undeveloped in the first decades of the 20th century, was promoted, but proselytism was focused on extending sexual education in institutionalized education and making it available for the rest of the population as well.

Aware of the fact that the prenuptial certificate and the scientific control of conception could only be directed to certain sectors of the population that were educated and fully aware of eugenic values, and that they would not solve the problem in its entirety, geneticists also waved, in a more radical stance, the flag of the sterilization of certain individuals or groups: feebleminded persons, cretins, criminals, etc. It was a quite extended practice in some countries but with different intensity, and even in those countries in which it was regulated by legislation, sterilization was not put into practice systematically. It was a common practice in the United States. In another case, the nazi law on "prevention of hereditary diseases" was promulgated on July 14, 1933, and established the several grounds for sterilization: "feeblemindedness, epilepsy, hereditary blindness or deafness, dementia praecox, schizophrenia, Saint Vitus's dance or convulsive disorder, serious malformation of the body and chronic alcoholism."

The objective of preventing the reproduction of certain groups of the population that were considered inferior led to other related practices in diverse countries. "The number of prefrontal lobotomies of different types carried out in the United States from 1936 to 1955 has been estimated between forty and fifty thousand" (Chorover 1979, p. 203). Systematic and lawful sterilizations were in their tens of thousands in the United States and had the same objective. The development of the programs of mandatory sterilization and mass extermination of human beings considered socially "undesirable" under nazism is history.

The implementation of eugenic abortion as a more extreme measure to exceed the limitations of birth control also was repeatedly proposed, even if an effective implementation was less frequent. In the specific case of abortion it should be pointed out that eugenicists did not defend it as a prerogative or an individual and voluntary decision of the mother. Accordingly, the intention was not to decriminalize abortion. On the contrary, the aim was to achieve an effective regulation in view of the fact that, as eugenicists affirmed, clandestine abortion was resorted to "by almost exclusively the individuals in our population that are capable of having better offspring [...]" (cited in Mac-Lean & Estenós 1952, p. 68).

In many countries, mostly Latin American countries with a majority of catholic population, eugenic abortion as well as birth control have encountered strong barriers originated by the clash between these measures and religious dogma. In this way, the Church, in spite of being usually connected with the sectors of society that are more conservative and even reactionary, operated in these cases as a limit for the excess that eugenic policies could produce. Whatever the case, eugenic abortion has been scarcely resorted to, probably because of the low level of efficiency and extent of this measure in connection with reproduction, if we compare it to other technologies implemented.

Another of the more widespread ideas has been the tendency to control, restrict or strongly intervene in the immigration of certain human groups, and even though the reasons given for adopting these measures have been quite heterogeneous in nature, eugenic considerations have been of a fundamental importance. As all the other measures, they were implemented differently according to the country receiving the immigrants: the American countries, Australia, some African countries, and the Balkans in Europe. However, it can be said that in all of them, the immigration policy followed a similar pattern in which two stages can be perceived. In the first stage, which, with some variations, extended over the first half of the 19th century, and in some countries much longer, policies were implemented in order to promote immigration. In a later stage, immigration was gradually restricted, not so much in quantity but rather by means of prohibitions against certain groups or "races". In all countries receiving immigration debates were held on the need to restrict the entrance of immigrants. Both the levels of prohibition and the acrimony of the debate have been variable but generally the purpose was to exclude blacks, Asians, gypsies, criminals, anarchists, madmen, alcoholics, syphilitics and persons suffering from tuberculosis.

There is a significant relationship between eugenics and education, which is stronger in those countries and sectors in which eugenics has taken on a profile not so definitely hereditarian but one in which environmental conditions are accepted as a major influence on individuals, their traits and behavior. It can be generally said that the relationship between eugenics and education presents two aspects. On the one hand, the conviction that the environmental aspects, mainly training and education among them, are fundamental because they can change the fate by which some individuals are doomed to degeneration and also because it was thought that the consciousness raised by information – basically on syphilis, alcoholism and tuberculosis – would prevent reproduction, or at least it would contribute to it not being dysgenic. In this context, it is not unusual to find strong claims for the implementation of sexual education, which is considered one of the cornerstones of depuration and improvement of a race. However, the sexual education proposed has always been connected with reproduction (or non reproduction) and to the responsibility in relation to the race, sexually transmitted diseases and alcoholism, that is to say, with a strong medical and biological bias. There are no references to the question of sexual pleasure except to the extent of considering it but to consider it as a kind of natural residue (and secondary) after the "natural" objective that reproduction is. The relation sex-reproduction becomes an issue and a sound reproduction is sought, but the relation sex-pleasure is never introduced. It is sought that the reproduction be regulated, rationalized and subjected to scientific control. Nonetheless, the struggle for introducing sexual education from the first years of school, even in the particular and biased sense in which eugenics was understood, has been tough and long. Another level on which eugenics and education are distinctly related, though including, in this case, an interrelation with school institutions is connected with the repeated claim by eugenicists for the control and classification of students – and all the population generally – through the so called "eugenic files".

A detailed analysis of the implementation of social policies and technologies in all the Western World shows, once again, the huge difficulty in distinguishing between positive and negative eugenics. On the other hand, the strongly technocratic aspect of eugenics appears emphasized because it mainly implies an increasing medicalization and/or "scientization" of the social relations and processes associated with a demand for the creation or strengthening of state institutions and the control over reproductive aspects through active policies or legislative reform that may convey a greater decision-making capacity to professionals and practitioners.

2.2. Medicalization: Eugenics as a technocratic practice

Probably one of the reasons for which eugenics enjoyed such esteem and was so influence and that also explains the demand for greater state regulation and administrative centralization of health policies is the fact that eugenics was rapidly perceived as an adequate response to the emergence of certain new problems in the areas of social medicine or public hygiene. A constant feature in all eugenic literature is the demand for the creation of institutions of various types directed to the control and follow-up of diverse pathologies and human groups. It was natural to consider that the State was responsible for the regulation of, among other matters, the process of human reproduction and that it had the power to restrict the reproduction of persons considered not fit, whether by means of education, stressing the fact that "the physical and mental condition of parents at the moment of conception" has great importance in relation to the biological constitution of children, or through more direct mechanisms such as the amendment of legislation on abortion to grant more freedom to physicians to make decisions on the matter. Generally, these measures were associated with programs of health reform in such areas as labor health conditions and the control, prevention and eradication of sexually transmitted diseases – basically syphilis –, as was the fight against alcoholism, prostitution and the unlawful use of drugs, considered by eugenicists, together with the associated tuberculosis, as the more severe manifestation of "racial poison".

In this scenario, the scientist that had the floor was the medical doctor, who, in turn, questioned the State and demanded its intervention. The medical doctor figure gradually appeared as the guarantor of the general well-being through the control of the individuals, which was brutally and dramatically legitimatized by the recurring epidemics that broke out in the last decades of the 19th century, the deficient health conditions of the large cities that manifested themselves in the increase of alcoholism, sexually transmitted diseases and tuberculosis, and the apparent "barbarism" of immigrants and, in some cases, of the native population. In this context, the doctor was not only a technician performing her or his specific work of healing but also an essential factor in civilization and progress.[7] This medicalization process presents two different and complementary aspects: on the one hand, the nearly unlimited but always indeterminate expansion of the sphere of influence of the medical science and doctors as a result of considering both what is "normal" and "pathologic" as basic categories of analysis; on the other hand, the demands for state intervention through diverse institutions and policies. These medical doctors, who not only healed the sick person but also the social organism and extended their intervention over new areas, now questioned the State and demanded of it preventive actions as well as control and enforcement measures in accordance with the diagnoses that they, as specialists, made. In addition, hereditary and environmental conditions could converge in these measures and the State was the one responsible for providing for minimal conditions of health in the en-

---

[7] In some countries like Argentina, it must also be recalled – and it is no trifling matter – that many medical doctors have held important governmental offices.

vironment. In the short and medium term, the objective was to provide assistance and to control and suppress the factors of race degeneration, and in the long term, to create a eugenic consciousness. There is a direct relationship with hygienism, a true association between medical ideals, science, the means of the State and the purity of race in relation to the affirmation of nationality.

The process of medicalization of society implied a series of diverse mechanisms that were gradually implemented: the creation of complete systems of information and registry of the health characteristics of populations, the designation of medical officials to supervise diverse regions, the organization of specific areas such as public hygiene to deal with the problems of, basically, the cities, establishing a true political scientific control of the environment and the control of individuals through the interplay of two aspects: first, to consider essential a categorization according to the dichotomy between normal condition and pathologic condition and second, to place the individuals in one of these categories. In this way, it could be considered that insanity, alcoholism, and the so-called sexually transmitted diseases, but also various sexual inclinations and conducts and criminality, were pathologies. Even the inclusion of individuals under the pathological categories resulted from procedures that were not methodologically defined and that could not pass a more or less strict epistemological examination. In the context of the considerations on these pathologies and the qualities of diverse groups or races regarded as inferior, the medical intervention developed through the treatment of sick persons but, to the extent in which such qualities were considered inherited, there were fundamental and increasing demands for "prevention", which gave the doctors a decisive intervention in reproduction and sexual conducts and, in the case of immigrants, a claim on entrance restrictions.

The argument for eugenics as an obligation of the State – which is the key to differentiate it from "liberal eugenics" – is grounded on the belief that society is the highest value and that it should prevail over individuals.

> There are duties to the family and those persons closer to us, duties to the State, to the existing mankind and to posterity. The last one is the highest duty of everyone. [...] Rationally speaking, a moral sys-

> tem must subordinate individual happiness to the happiness of the community in general. (Forel 1912, p. 661)

> That supreme law that is the health of the people comes before individual interests, and in its name the lawmaker must make use of all his authority in order to pass statutes embodying it, without paying any heed to the considerations of those who theorize on a hypothetical freedom that stealthily forges chains. (Farré 1919, p. 94)

The concept of "social defense", that is interwoven with the consideration of "public order" as an essential value, is the key to understand the legitimacy of the demand for different actions that the State should carry out. Preserving public order and social defense are primary aspects that manifest themselves in the ideals of purity of the race, specific health measures, and also in the considerations on new sources of legitimacy of criminal punishment – without focusing on the responsibility of criminals but on the defense of society –, in restrictions on immigrants considered undesirable, the elimination or commitment of lunatics, and even the formulation of a "sexual ethic".

The relationship that we have pointed out is particularly evident in the debates that have taken place in the field of criminology since the end of the 19th century. In fact, the close relationship between eugenics and criminology is detected in different ways, mostly from the developments started by the Italian criminological school, with Cesare Lombroso, a medical doctor, as its best known figure. Towards the middle of the 19th century, the old pre-Darwinian idea of recapitulation regained strength in the midst of a euphoric evolutionist climate. As handled by Lombroso, it gives rise to the theory of the born criminal, according to which criminals, from the point of view of evolution, are atavistic types that persist in human beings. According to Lombroso, dormant germs lie in human inheritance from a primitive past, which reemerges in some ill-fated individuals that are driven by their innate constitution to behave like common monkeys or savages would; however, in our society their conduct is considered criminal. Lombroso held that, fortunately, born criminals can be identified because their apelike character manifests itself in certain anatomical traits. Their atavism is both physical and mental, but the physical traits, or "stigmata", are conclusive. A criminal conduct can also appear in normal men, but the "born criminal" is recognized by his anatomy. Without taking into account the heresies that would later diverge to a greater or lesser extent from the original version, the Lombrosian theory established, over decades, the basic agenda on the treatment of crime, placing discussions on and devices of detection and control outside the specifically social and cultural dimension and changing the focus of attention on the internal individual psychophysical organization, practically nearly always focusing on the lower social conditions. Lombroso established a true typology of criminals out of the measures of the different parts of the body, for example the length of the arms and cranial size, or traits such as facial asymmetry and other facial characteristics. He determined a large number of ape-like stigmata, which denoted innate criminality: a thicker cranium, simplicity of cranial sutures, large jaws, precocious wrinkles, a narrow low forehead, big ears, lack of baldness, darker skin, sharper eyes, low sensitivity to pain, lack of vascular reaction (inability to blush). These theories have had a huge influence on criminology and international legal literature and not only as an academic debate but as a part of criminal law practice. Lombroso's influence generated a new way of conceiving punishment. While the classical school of criminal law considered that the punishment should exactly correspond with the nature of the crime, Lombroso held that it should correspond with the criminal. Hence, Lombroso's object of study was not the crime but the criminal. Once the criminal was identified, the punishment administered was not grounded on the individual responsibility of the person that committed the crime but on the need of the community to "defend itself". Thus, it was legitimate to punish a born criminal because of the commission of a lesser offense due to the fact that, inevitably, he would do it again and consequently, it was ineffective to insist on his rehabilitation. As a counterpart, it was rather useless to punish an occasional criminal that would not commit a crime again. Therefore, the rationale behind a sentence would be a requisite for social defense rather than a punishment for the criminal, who ultimately was a sick person. Ferri, a follower of Lombroso, similarly held the concept of the "indeterminate sentence", that is to say that punishment should correspond to the criminal personality, even though classical criminologists regarded it as heresy. Sentences determined beforehand would be absurd from the point of view of the defense of society. It is interesting to point out that Lombroso's ideas admit of a double application: on the one hand, they ideologically stigmatize the alleged criminals and on the other hand, they furnish reasons for mitigating sentences on the basis of the "natural" character of the "criminal instinct". This way, some later Lombrosians that expanded the determination of criminals to include environmental factors such as education contributed to establish the idea of the mitigation of sentences, because of mitigating circumstances.

The Lombrosian school expanded and introduced substantial modifications to the initial formulations, mostly in regard to atavistic traits, but started a model of conceptualizing and detecting criminals that has lasted for decades, based on the idea that criminality manifests itself in some particular organic constitution. The extensive developments of the diverse versions of biotypology[8] are recipients of these basic ideas that, in the context of the concern for the biological constitution of the population and the "improvement of the race", contributed to strengthening the intricate fabric of ideas that "substantiated" the superiority of certain racial groups above others, for example the inclusion of anarchists and working-class fighters as a case of biological criminality. These ideas also contributed to including criminality in the process that we have called "medicalization" or "biologization": the explanation for and solution to the problem of criminality was a concern of medicine and psychiatry, and the whole criminal and legal system should be subsidiary to them. A good number of criminals were considered "sick". To a large extent, a trilogy illness-insanity-crime is established that feeds itself back and justifies itself in an immanent manner. Besides, one of the aspects shown by the demands for restrictions on immigration is related to criminality and the criminalization of working-class fights, mainly the immigrants that were anarchists. An increasing correspondence between criminality and immigration gradually built up, correspondence that is ultimately rooted in a more basic one: the correlation between race and

---

[8] "Criminal biotypology is the study of the inherited characteristics, the morphological habit, the humoral dynamic temperament, the character and intelligence, that is to say the entire psychosomatic personality of the criminal, in order to determine the individual biotypological characteristics of criminals and set out the true scientific provisions in statutes and police regulations, which involves the punishment or correction for those who committed a crime [...]" (Bosch *et al.* 1934, p. 8).

crime. Gradually, anarchism ceased to be a problem or a social issue and almost exclusively became a part of the general process of criminalization that, according to common beliefs, emerged as though it was undergoing an uncontrollable process of augmentation on which the efforts of the State should focus.

2.3. "Racialism": Eugenics as an authoritarian practice

Racist considerations as a cultural feature of mistrust and segregation of what is strange are probably as old as humanity; however, the 19th century shaped an approach that consisted in finding some scientific justification for the differences and hierarchies between the races, which justification has nearly always operated – this is the well-known history of racialism in the 19th and 20th centuries – as a self-fulfilling prophesy, that is to say, it always sanctioned and legitimized existing inequalities.

Count Joseph Arthur of Gobineau (1816-1882) has traditionally been identified as one of the greatest exponents of 19th century racialism since the publication of his *Essay on the Inequality of Human Races* in 1853-55, in which he expounded a universal history based on the characteristics of human races – Negroid, yellow and white –, exalting the virtues of the white race as the best exponent of human values: energy, intelligence, love of life, productive and speculative capacity. Nevertheless, according to Count of Gobineau the mixing of races is prejudicial to humanity because of the fact that, even though inferior races profit from miscegenation, superior races lose on the mixing more than they win, and therefore, the result is negative. In the opinion of Gobineau, a thinker imbued with a romantic, aristocratic and colonialistic mentality, history is the result of the hegemony of superior races, and he considered that the purity of blood – in his case, French blood – should be preserved at all costs in order to prevent the degeneration of humanity. Whatever the case, racialism, presenting different and sometimes significant nuances and also contributing with clearly differentiated proposals on the legitimized practices and the resulting social technologies, was a largely extended view that, over the last decades of the 19th century and the first decades of the 20th century, influenced almost every academic, scientific and political sphere. It was almost commonplace in all scientific literature to point out that there were inferior races (blacks, Polynesians and

Asians) and superior races; on the basis of a combination of craniometrical traits, cultural prejudices and sociological considerations it was sought to sanction such hierarchy. Within this general framework, eugenics installed itself in accordance with the consideration that there are human groups that are to be favored because they are superior or better, added to which is the complementary purpose of preventing the decadence of the species/race that results from the commiseration and protection that some societies give these inferior groups, which, besides, reproduce themselves in larger quantities. Although what we have described is the more standardized version of racialism as associated with eugenics, some more considerations are in order here.

Even though the most typical way to distinguish one race from the other is anatomical, it is also possible to find other complementary and even contradictory ways to do it: on some occasions, biological factors were included; on some other occasions, the factors were geographical, climatic, historical and cultural; sometimes race was confused with nationality, or the sum of biological and cultural characteristics were considered as a whole. In this sense, plentiful eugenic literature can be found, in which the concept of race may almost be synonymous with the concept of "population", mostly when writers that adopted the eugenic discourse took softer lines that were only related to the eradication of some diseases.

A second important aspect to bear in mind is that the features that are common to all these conceptions are based on the establishment of hierarchies and the implementation of measures to secure them or arrange them; however, even though the most common version points out the superiority of the white race, it only provides the starting point from which the discussions grow in subtlety. This way, it is common enough to find that distinctions between superiors and inferiors are made even within the white race, generally associated with differences in social class.

In the third place, the concept of superiority/inferiority that stems from the concept of race in many cases supports the idea that a race has to be "pure"; this "purity" has to be recovered or maintained. But in other cases what is defended is that the "mixture of races" is better and then, disputes revolve around what the most desirable mixture is.

Lastly, the racialist framework and the immigration policies of many countries are closely interrelated and, as seems logical to assume, in those countries which expelled a part their population, basically European countries, eugenics is directed to the preservation of the purity of the race, while in the countries receiving immigration and in which the native population has been supplanted the discussions revolve around what immigration should be admitted in order to achieve an adequate racial mixture. The Argentine case is typical in this sense.

3. Can a New Eugenics be expected?

Finally, we will try to answer the two questions posed at the beginning of this paper. Firstly, in what sense can we speak about a new eugenics? If we describe eugenics most generally as any manipulation of offspring, it seems possible to liken the concept to the term, but at first this approach does not seem interesting and besides, according to what can be inferred from the descriptions presented above, it seems a mistake from an historical point of view. It is not interesting because it prevents the elucidation of the complexity of the problem or, in any case, the multiple perspectives that the question involves. And it is a mistake from an historical point of view that prevents the comprehension of the scope and the implications of the eugenic world movement. Individual decisions freely made with the intention of determining some traits of the offspring – whether to eliminate serious illnesses or to attain a desired trait – can undoubtedly have significant implications for the life of the offspring, its family and close friends, but at first it does not much resemble the implementation of public policies that are coercively exercised with the purpose of selecting definite groups to influence the future evolution of the species. As affirmed above, eugenics as a practice is technocratic and authoritarian. It would be a contradiction in terms to argue for the existence of a "liberal eugenics". In any case, we have to be careful when we speak about the choice of desirable characters because, even if they cannot be explicitly imposed by the public power, these characters can be chosen out of racial or gender prejudices and in this sense, the choice, even on an individual decision-making level, is based on and strengthens the social nature of prejudice.

Secondly, we have to answer whether eugenic practices can be reinstated. It is always risky to say something about the future responsibly, but if we confine ourselves to the growing technological developments and the deterioration of the political conditions of democracies, it is not unthinkable that, in the future, eugenic programs both in the negative and the positive sense may be implemented. Undoubtedly, technological capabilities are more effective than a century ago and will be even more effective in the future. However, it is probable that such eugenic programs, even though in the past they have occurred in democracies of a liberal nature, can only be carried out in a society of an Orwellian type, in which fear aroused by a powerful system of propaganda causes a large part of the population to view as possible – and even desirable – the loss of some of its civil rights with a vision of achieving a society determined by values of a biological character. Even if this political scenario did not happen and consequently it were possible to achieve an unrestricted exercise of the fundamental rights that guarantee the self-determination of citizens, it could then happen that by stirring up the ghost of eugenics, with all the dreadful historical burden associated with it, the concealment of other social practices would be favored – subtler social practices, which many times are not perceived as a violation of fundamental civil rights but as a necessary interplay of market forces and which find in modern genetics and the medicalization of the social relationships a framework within which they can be justified. Two examples from present times can be eloquent enough: the exclusion from or limitation in medical insurance and coverage and in access to jobs for those carrying certain genetic alleles. Therefore, it is usually demanded that certain actual or imagined practices deriving from the knowledge in the field of cell genetics should be subject to moral and legal brakes based on the individual legal guarantees and the human rights that Modernity has gained with much suffering, even though it is not clear how to carry out these demands in the first place. At least, we have to guarantee that a public, continual and well-grounded debate may be held, which is certainly hindered by the growth of poverty, the marginality in the world and the arguments for security and censorship that grow in the unipolar world in the name of war against terrorism. It has to be clearly stated that the best way never is to choose ignorance. Those that dream that, by stalling scientific development, they would prevent the emergence of the challenges posed by modern genetics forget two questions: the first one is that the eugenic movement unfolded in spite of being founded on theoretical assumptions that were rather weak. If we were to fear the return of eugenics as formulated in the past, it would be probably based on at least two problematic real situations: overpopulation and the new immigration currents. But this line of argument of eugenics that will be attempted to be legitimized as scientific will show again the paradox of eugenics being judged by its adaptation to certain ideological lines and eugenics not recognizing its ineffectiveness and the legitimate criticism leveled from the very field of science. The second question entails the admission that science is not autonomous and that, even if it presents an internal logic, it is a part of culture and, as such, is submerged in the conflict of ideas in which the men of that same culture are engaged.

References

Andorno, R. (2001), "El derecho frente a la nueva eugenesia: la selección de embriones *in vitro*", available in: <http://cuadernos.bioetica.org/doctrina-htm>.

Bernal, J. (1954), *Science in History*, London: C.A.Watts and Co. Ltd.

Beruti, J. & A. Rossi (1934), "Ficha eugénica de valuación de le fecundidad individual", *Anales de Biotipología, Eugenesia y medicina social* 2 (30): 12-17.

Bosch, G. (1930), "Los propósitos de la 'Liga argentina de Higiene Mental", *Revista de la Liga Argentina de Higiene Mental* 1 (1): 4-10.

Bosch, G., Rossi, A. & M. Rodríguez (1934), "Biotipología criminal", *Anales de Biotipología, Eugenesia y Medicina Social* 2 (30): 7-11.

Chorover, S. L. (1979), *From Genesis to Genocide*, New York: MIT Press.

Figueroa, F. (1906), *Las huelgas en la República Argentina y el modo de combatirlas*, Buenos Aires: Imprenta Tragant.

Forel, A. (1912), "Etica sexual", *La Semana Médica* 19 (40): 666-668.

Frers, E. (1918), "La inmigración después de la guerra", *Boletín del Museo Social Argentino* 6: 1-186.

Galton, F. (1883), *Inquiries Into Human Faculty and Its Development*, London: Macmillan.

Galton, F. (1884), *Hereditary Genius*, New York: D. Appleton.

Gould, S.J. (1981), *The Mismeasure of Man*, New York: Norton.

Grasa Hernández, R. (1993), *El evolucionismo: de Darwin a la sociobiología*, Madrid: Editorial Cincel.

Habermas, J. (2001), *Die Zukunft der menschlichen Natur. Auf dem Weg zu einer liberalen Eugenik?*, Frankfurt/am Main: Suhrkamp.

Ingenieros J. (1904), "Evolución de la antropología criminal", *La Semana Médica*: 1374-1380.

Hobsbawn, E. (1987), *The Age of Empire 1875-1914*, London: Weidenfeld and Nicolson.

Ingenieros, J. (1915), "La formación de la raza argentina", *Revista de filosofía, cultura, ciencias y educación* 2 (6): 464-483.

Ingenieros, J. (1927), "Amor, intereses y eugenesia", *Revista de filosofía, cultura, ciencias y educación* 16 (4): 110-132.

Kehl, R. (1926), "La eugénica y sus fines", *La Semana Médica* 33: 479-481.

Kevles, D.J. (1995), *In the Name of Eugenics*, Cambridge, MA: Harvard University Press.

Lafora, G. (1931), "La esterilización eugenésica de los degenerados", *Boletín del Museo Social Argentino* 19: 360-363.

Lafora, G. (1931a), "Eugenesia", *Boletín del Museo Social Argentino* 19: 92-94.

Luján López, J. (1996), "Teorías de la inteligencia y tecnologías sociales", in González García, M. (ed.), *Ciencia, tecnología y sociedad*, Madrid: Editorial Tecnos, 1996.

Mac-Lean y Estenós, R., (1952), *La eugenesia en América*, México: Instituto de Investigaciones Sociales-Universidad Nacional Autónoma de México.

Massin, B. (1991), "Del Eugenismo a la «Operación eutanasia» 1890-1945", *Mundo Científico* 110 (11): 206-212.

Maynard-Smith, J. (1982), "Eugenesia y utopía", in Manuel, F.E. (ed.), *Utopías y pensamiento utópico*, Madrid: Espasa Calpe, 1982, pp. 194-214.

Melcior Farre, V. (1919), "Degeneración y regeneración de la raza", *La Semana Médica* 26 (30): 77-99.

Montagu, A. (1980), *Proceso a la sociobiología*, Buenos Aires: Editorial Tres Tiempos.

Müller-Hill, B. (1984), *Tödliche Wissenschaft*, Reinbek bei Hamburg: Rowholt.

Paul, D. (1998), *The Politics of Heredity. Essays on Eugenics, Biomedicine and the Nature-Nurture Debate*, New York: State University of New York Press.

Palma, H. (2002), *"Gobernar es seleccionar". Apuntes sobre la eugenesia*, Buenos Aires: J. Baudino Ediciones.

Pende, N. (1935), "Biología de las razas y unidad espiritual mediterránea", *Anales de Biotipología, Eugenesia y Medicina Social* 3 (41): 12-18.

Plotkin, M.B. (1996), "Psicoanálisis y política: la recepción que tuvo el psicoanálisis en Buenos Aires (1910-1943)", *Redes* 3 (8): 163-198.

Randall, J. (1926-1940), *The Making of the Modern Mind*, Boston: Houghton Mifflin.

Regnault, J. (1922), "La eugénica", *La Semana Médica* 29: 22-25.

Ruse, M. (1973), *The Philosophy of Biology*, New York: Hutchinson and Co.

Ruse, M. (1980), *Sociobiology: Sense or Nonsense?*, Dordrecht: Reidel.

Simpson, G.G. (1951), *The Meaning of Evolution*, New Haven: Yale University Press.

Soler, R. (1968), *El positivismo argentino*, Buenos Aires: Paidós.

Soutullo, D. (2001), "Actualidad de la eugenesia: intervenciones en la línea germinal", available in: <http:// www. ugr. es/ ̄eianez/ Biotecnologia/eugenesia.htm>.

Stach, F. (1916), "La defensa social y la inmigración", *Boletín del Museo Social Argentino*: 361-389.

Stepan, N.L. (1991), *The Hour of Eugenics: Race, Gender and Nation in Latin American*, Ithaca: Cornell University Press.

Taylor, H. (1980), *The IQ Game: A Methodological Inquiry into the Heredity-Environment Controversy*, New Jersey: Rutgers University Press.

Testart, J. & Ch. Godin (2001), *Au bazar du vivant*, Paris: Éditions du Seuil.

Thuillier, P. (1988), *Les passions du savoir. Essais sur le dimensions culturelles de la science*, Paris: Librairie Arthème Fayard.

Todorov, T. (1989), *Nous et les autres. La réflexion française sur la diversité humaine*, Paris: Éditions du Seuil.

Vezzetti, H. (1985), *La locura en la Argentina*, Buenos Aires: Paidós.

Wilson, E. & C. Lumsden (1975), *El fuego de Prometeo*, México: Fondo de Cultura Económica.

Wilson, E.O. (1975), *Sociobiology. The New Synthesis*, Cambridge, MA: Harvard University Press.

Zimmermann, E. (1995), *Los reformistas liberales*, Buenos Aires: Editorial Sudamericana-San Andrés.

The Emergence of a Research Programme in Genetics[*]

Pablo Lorenzano

National University of Quilmes (UNQ)/
National Scientific and Technical Research Council (CONICET)

1. Introduction

According to what might be called the "official story of genetics",[1] it came to life as a discipline in 1865 when the Austrian monk Gregor Mendel made public the results of his experiments with peas in the Natural Science Society in Brno, results that would be published in the *Proceedings* of such society a year later. However, his work remains either unknown for the most part or, when that was not the case, misunderstood until it is rediscovered in the year 1900 simultaneously and independently by three researchers (Hugo de Vries in Holland, Carl Correns in Germany and Erich Tschermak in Austria) who,

---

[*] This paper was written with the aid of grants conceded by Antorchas Foundation and the National Agency of Scientific and Technological Promotion (ANPCyT), Argentine.
[1] The preference for this denomination instead of those used by other authors such as "traditional account" (Olby 1979) or "orthodox image" (Bowler 1989) shows a clear influence of Argentinian cinema: "The official story" is the title of the Argentinian film that was awarded the Oscar as best foreign film in 1985. [Note: The single Spanish term "historia" stands for both story and history, hence the play of words.]

working on the same problem, independently obtain the same results as Mendel – the proportions 3:1 and 9:3:3:1 and their explanation by means of the law of segregation and the law of independent transmission. Meanwhile, William Bateson reads Mendel's paper in England, recognizes its importance immediately and starts publicizing it, so that Mendel comes to be honored as the father of genetics and has a place thus secured for him in the history of science. Ten years later, Thomas Hunt Morgan and his disciples start basic research in genetics and by relating it with the available knowledge of cytology, they investigate and explain the apparent exceptions, widen its field of application and so help to give form to the so called "formal", "classical" or "Mendelian" genetics, such as the universally acknowledged theory of heredity.

In other places and in line with the new historiografy of genetics I have tried to show that this story is nothing but a myth.[2] Here I will hold that genetics crystallized as a distinct biological discipline with difficulty through the work of William Bateson and his collaborators. This happened neither overnight nor without opposition. On the contrary, this is a process that took place during a great part of the first decade of the twentieth century and in which Bateson's so called "Mendelism" had to take stand against other perspectives that addressed the problem of heredity at the time, such as biometrics, cytology and experimental embryology. However, despite the lack of *complete agreement* on the part of the scientific community *either before or after* such crystallization on the issues of what were the problems to be solved, what were the acceptable answers, what were the criteria that such answers should satisfy, what were the appropriate techniques and what were the interesting phenomena, it was the research programme founded by Bateson the one that would come to bear the name "genetics" and would have the greatest acceptance by the scientific community early in the second decade of the twentieth century.[3] This research programme, on the other hand, differs both from the work done by Mendel and his rediscoverers and from that carried out later by Morgan and his collaborators.

---

[2] See Lorenzano (1995, 1997, 1998a, 1999, 2000b) and the bilbiography mentioned there.

[3] About the process of building the consensus in such community, see Kim (1994).

This paper has a double purpose. First, it seeks to expose some of the conceptual and methodological changes that took place within the study of the problems of heredity during the first decade of the twentieth century and lead to the appearance of the first defined research programme in genetics[4] through the theoretical developments known by the name of "Mendelism"[5] due to Bateson and his collaborators. Second, it seeks to characterize such programme.

2. Delimitation of the central problem

One of the changes that occurred during the first decade of the twentieth century is related to the central problem to be addressed and the consequent restriction of the field of application. The theories of heredity elaborated in the nineteenth century had very wide domains of application: they were often not only theories of heredity – that tried to explain why offspring resembled their parents –, but also theories of embryological development, normal growth and evolutionary change. Some twentieth-century biologists also wished for a unified theory that could account for evolutionary change, inheritance, variation, sex determination and embryological development.[6] Bateson himself, starting in 1883 with some research on morphology and embryology, focuses on a rigorous study of variation and heredity in order to give an answer to what was for him the central problem in

---

[4] On the role played by the Royal Horticulture Society both in the emergence of "Mendelism" and in the support of the investigations done by Bateson and his collaborators and the consequent securing of its position, see Olby (2002).

[5] Although this denomination first appeared in 1903 to refer to the works done by Bateson and other "Mendelians" before the foundation of genetics as an autonomous and independent discipline (see, for instance, Bailey 1903), it continued to be used afterwards to refer specially to the theoretical developments due to Bateson and his collaborators, which, according to our interpretation, constitute the first research programme in genetics (Reginald C. Punnett's book *Mendelism* constitutes its first systematic exposition, it was first published in 1905 and re-edited with successive modifications in 1907, 1911, 1912, 1919, 1922, 1927). Among Bateson's collaborators in the period we should count, besides his wife Beatrice, E.R. Saunders, R.C. Punnett, J.B.J. Sollas, H. Kilby, Wheldale, Marryat, F.M. Durham, R. Staples-Browne, L. Doncaster, R.P. Gregory, R.H. Lock, R.H. Biffen and C.C. Hurst.

[6] For an account of the multipliciy of theories of heredity that existed early in the twentieth century, see Delage (1908).

biology: the problem of evolution. And it is in this context that he carries out his research.[1] However, there is a shift of the core of his main activities that can be observed in his work: first, from evolution to variation, and later, from variation to heredity. This does not mean that Bateson neglected completely the first of his problems. Instead, what happened was that the questions of variation and heredity grow gradually independent from the evolutionary issue, to the extent that they become a new field, a new discipline. In the year 1906 Bateson distinguishes both problems clearly and explicitly, and acknowledges the issue of variation and heredity to have its own identity and independence,[2] giving the name "genetics" to the branch of biology that is concerned with it.[3] This does not mean that both fields are not related, but that the relationship between them is no more an *intra*disciplinary one but an *inter*disciplinary one.

There is an analogous situation in the case of embryology. Early in the twentieth century the problem of transmission of hereditary traits was not clearly separated from the problem of explaining how those traits developed during ontogeny. Even though Bateson always kept a strong interest in embryology and the desire to find a conceptualization that could comprehend both issues – he held, for instance, that the existing analogies between heredity and development were so strong that they could be said to be "phenomena of the same kind"[4] – he concentrated on the study of the transmission of traits, developing a theory that accounted for it and did not include – though it was related to – embryology.

---

[1] It is also in this context where his debate with "biometrists" W.F.R. Weldon and K. Pearson is situated. About this debate, see Provine (1971), Froggatt & Nevin (1971a, b), Cock (1973), Norton (1973), Marrais (1974), Farrall (1975), Norton (1975a, b, 1978), MacKenzie (1978, 1979), MacKenzie & Barnes (1975, 1979), Roll-Hansen (1980, 1989), Olby (1988) and Kim (1994).
[2] See Bateson (1906b).
[3] The term "genetics" was used for the first time by Bateson in a letter to the Cambridge zoologist Adam Sedwick in 1903 (Bateson 1905a) and publicly in 1906 (Bateson 1906a), as a result of which the "Third Conference on Hybridisation and Plant-Breeding" was renamed for publication of its proceedings as "Third International Conference on Genetics".
[4] Bateson (1909, pp. 274-278). For Bateson's embryological background, see also Bateson (1894, pp. 254-257), and Bateson (1901b, pp. 404-405).

2. Bateson's theory: "Mendelism"

The road to theoretical clarity was not easy in genetics, nor was it attained immediately either, but only in the course of time and through constant proposals, counter proposals, discussions and modifications that took place within the scientific community. And though Bateson had a different view to the one described in current textbooks under the name of "classical genetics", he soon contributed much to that process and around 1905 developed a theory of heredity based on factors – "Mendelism" –, that was known as "genetics" until the appearance of the theory due to Morgan and his disciples.

Early in the twentieth century it was not customary to separate explicitly what is now known as "law of independent transmission", or "Mendel's second law", from the "law of segregation", or "Mendel's first law". In fact, and contrary to what is usually claimed, we cannot find in Mendel a statement of the laws attributed to him in the terms they are usually presented.[5] It was Hugo de Vries (1900b, c) the first to speak at the dawn of the twentieth century of the "law of segregation of hybrids" (*"loi de disjonction des hybrides"* in French and *"Spaltungsgesetz der Bastarde"* in German) and to acknowledge Mendel as its discoverer. Carl Correns (1900a), another of the "rediscoverers", uses the expression "Mendel's rule" (*"Mendels Regel"* in German) at the time to refer both to de Vries' "law of segregation" and to what came to be known later as "Mendel's second law". The use of the term "rule" rather than "law" is due to the fact that Correns preserves this latter term to refer to statements that, in contrast with rules, posses universal validity (Correns 1900b). The first to use the term "independent transmission" was Thomas Hunt Morgan (1913), who was also the first to speak explicitly of two laws, the law of segregation and the law of independent transmission of genes, attributing its discovery to Mendel and referring to them thus as "Mendel's first law" and "Mendel's second law", respectively (Morgan 1916).

---

[5] For an analysis of the "law found in *Pisum*" "concerning the formation and development of hybrids" – *für die Bildung und Entwicklung der Hybriden* –, consisting of the "law of simple combination of characters" –*Gesetz der einfachen Kombinierung der Merkmale* – and the "law of combination of differential characters" – *Gesetz der Kombinierung der differierenden Merkmale* –, as well as other aspects of Mendel's work, see Lorenzano (1997).

In the writings that addressed the problem of heredity early in the twentieth century, also, it was not easy to recognize a clear and explicit distinction between external hereditary characters and the "something" responsible for them that was transmitted from one generation to the other. De Vries, for example, referring to the "law of segregation of hybrids", spoke of the "segregation of characters" – "*caractères*" in French and "*Merkmale*" in German – and not of "factors" or "genes". Carl Correns is an exception in this sense, because in 1900 postulates explicitly a hereditary unit or *Anlage* (following terminology from his teacher and Mendel's correspondent, Carl von Nägeli 1884) for each character in individuals and claims that these are always found in pairs in somatic cells.

Bateson, for his part, formulates originally the law of segregation in terms of types of gametes and not of hereditary units present in the gametes, and holds that the essence of Mendel's view lies in the purity of the gametes. The fact that germinal cells – called "gametes" – establish the connection between the different generations, being the material transmitted from parents to offspring, is a part of cytology and is presupposed in the theory of – sexual – reproduction, and was used by early Mendelians in the formulation of the law of segregation, without distinguishing between the genetic processes, on one hand, and the cytological processes that support them on the other.

Another change introduced by Bateson in the study of the problem of heredity – now posed independently of questions regarding evolution and the ontogenetic development and restrictedly as the problem of the transmission of hereditary characters – consists of the formulation of his "factorial hypothesis", originated around 1905.[6] It allows us to appreciate the novelty of Bateson's proposal, as well as his efforts, unique during that first decade of the twentieth century, towards precision and conceptual clearness.[7]

---

[6] Bateson & Punnett (1905), Bateson, Saunders & Punnett (1905), Punnett (1905).

[7] Appart from the already mentioned expressions "genetics" and "purity of the gametes", Bateson introduced the terms "allelomorph", "gamete coupling", "heterozygote", homozygote", "epistatic", "hypostatic" and "factor" – this last one for hereditary units –, while the denominations P for progenitors and $F_1, F_2, F_3...$ for the first, second, third and the rest of the filial generations – that is, of offspring – were also proposed by him (following Galton and not Mendel) and became universally accepted.

According to such conception, characters are not literally transmitted by the gametes. The ones responsible for the transmission and consequent appearance of certain features or characters are certain elements or units, called "unit-characters" first and "factors" afterwards, that are transmitted from parents to offspring through germinal cells or gametes during fecundation. Such factors are found in pairs in the individual (called "allelomorphs"[8] and are obtained one per each progenitor), while during the formation of gametes they separate ("segregate"), thus finding only one allelomorphic factor per gamete. In this last process there is a clear distinction between hereditary characters, on one hand, and hereditary units or factors responsible for such characters, on the other,[9] even when their nature (material or otherwise)[10] remains unknown.

The factorial hypothesis was in Bateson associated from the start to another hypothesis, characteristic of Mendelism, denominated "of *presence-and-absence*", according to which, the only two possible states

---

[8] With the help of the concepts "character" and "unit-character", he introduces, in 1902, the terms "homozygote" and "heterozygote" for those individuals that posses in the zygote two gametes either of the same type (with the same unit-character) or of different type (with different unit-characters). The expression "allelomorph" – later abbreviated by Morgan and his collaborators as "allele", designating the alternative states of a gene – was introduced originally to refer to a pair of observable differential unit-characters. From 1905 on two individuals are considered "homozygotes" when they posses allelomorphs of the same type in the zygote, and they are considered "heterozygotes" when they posses different allelomorphs.

[9] The terms "gene", "genotype" and "phenotype" are introduced by Johanssen (1909) with a slightly different meaning from the one they would acquire in the hands of Morgan and his collaborators, i.e., in the so called "classical genetics". About this, see, besides Johanssen (1909), Johanssen (1911, 1923), Churchill (1974), Wanscher (1975) and Roll-Hansen (1978).

[10] Coleman (1970) holds that Bateson's preferences were on the side of the non-material nature of these postulated entities. In support of that interpretation, see Bateson (1928, pp. 39-46; 1902, p. 274 ss.; 1917; 1913, chaps. 2 y 3; 1916, p. 462), where he conceives of them as dynamic non-material entities of the sort of "forces" or "vortices". Van Balen (1986, 1987) claims even that this position regarding the ontological status of such entities constitutes a feature or constraint of "Mendelism" (considered as research programme). On our part, however, we believe that this is taking things too far, because, for example, the preferences of Bateson's closest collaborator and unchallenged "Mendelian", Reginald C. Punnett, were not on this side (see, *e.g.*, Punnett 1907, p. 24).

of any factor present in the gamete are either its presence or its absence. When the factor is present, the character determined by it is manifested; when the factor is absent some other previously hidden character is likely to manifest. Thus, if the factor for yellow color, for example, is present in a pea, the seeds are yellow, whereas if it is absent, the seeds are green. The green character is there underlying all yellow seeds, but it can only be manifested in the absence of the factor for yellow color, and green color is allelomorphic to yellow, because it is the expression of the absence of yellow. The factorial hypothesis, in the interpretation provided by the presence-and-absence hypothesis explains without effort the 3:1 proportions of monohybrid crossings. And if it is supposed that factors are inherited completely independently from each other, it can also explain easily the 9:3:3:1, 27:9:9:9:3:3:3:1, etc. proportions of dihybrid, trihybrid, etc. crossings.

3. Bateson and the belief in the "promise" of Mendelism

At the same time, Bateson develops the conceptual framework known as "Mendelism", he broadens its field of application and develops an increasing trust in such conceptualization and its "promise"[11] of a fruitful research work.

At the turn of the century Bateson was not convinced of any of the available theories of heredity; while he acknowledged Galton's law of ancestral heredity to have applications,[12] he considered the extent of its validity an open question.[13]

In 1900, when Bateson reads one of de Vries' "rediscovering" papers and comes to know the law of segregation, he finds that Galton's law of ancestral heredity cannot be applied to all the cases that

---

[11] The first to use this expression in the analysis of Bateson's work was Darden (1977).
[12] When Bateson refers to this law, he does so having in mind not Pearson's reformulated version, but Francis Galton's original form, according to which, if hereditary material, the "hereditary mass", as a whole, is considered, then the contribution of both progenitors to such hereditary mass or to the hereditary properties of the offspring ammounts to ½, that of the 4 grandparents ¼, that of the 8 grandgrandparents □, etc. in such a way that the contribution of the totality of ancestors to hereditary properties of individuals can be expressed by the series ½ + ¼ + □ + …(½)^n.
[13] Masters (1900), Bateson (1900b, p. 174).

show dominance; at least not in the manner it was usually presented. Such cases are explained with the help of de Vries' law of segregation, that requires for its articulation – as Bateson claims at the time – only a modification of Galton's law.[14] Bateson accepts thus two hereditary patterns – one for mixed heredity, that shows no dominance and follows Galton's law, and the other for non-mixed heredity, that shows dominance and obeys Mendel's law –[15] believing that both are compatible and that Mendel's law is subordinate to Galton's. This was supposed to achieve an extension of the field of application of this latter law.

Two years later, contrary to his previous thoughts, Bateson came to think that the fore mentioned laws were not theoretically compatible, and, therefore, that the question was no longer that of deciding which one of them was subordinate to the other, but rather that of determining to what extent what had been accepted so far as the field of application of one of such laws – Galton's – should be considered in fact as belonging to the other law – Mendel's.[16] So, from 1900 to 1902, Bateson modifies his own opinion on the relationship between Galton's and Mendel's laws and the corresponding areas in which they are valid.[17]

For that purpose he started distinguishing between the discontinuous state of the gametes and the discontinuity exhibited by the individuals originated by them. Assuming then that the mixture observed in the characters that were not in a dominant or recessive relation did not require a corresponding mixture in the hereditary forms of their gametes, lead him to reject the restriction of Mendelian heredity to unmixed characters[18] and to sketch Mendelian patterns – developed later by others – for continuous characters.[19]

---

[14] See Maxwell Masters' (1900) report on the lecture given by Bateson on May 8th; on the moment Bateson takes notice of the law of segregation, see Olby (1987).
[15] Bateson (1900b, pp. 177-178).
[16] To see how Bateson changed his conception, see Bateson & Saunders (1902) and Bateson (1902).
[17] Olby (1987).
[18] Bateson & Saunders (1902, p. 59).
[19] Though he was aware of the difficulties of its testing (Bateson & Saunders 1902, p. 60). Bateson had speculated already in 1902 that four or five pairs of allelomorphs could act jointly and produce this way "continuous" characters. Pearson

However, and notwithstanding the fore mentioned extension of its field of application, Bateson does not attribute universal validity to Mendel's law,[20] although he does believe that the conceptual framework where it belongs makes possible, through certain alterations and theoretical developments, the progressive addition of more and more applications, not only within the vegetal realm but also, jointly with the Frenchman Cuénot (1902), to the animal realm.

It is in the light of such conviction that Bateson's treatment of some exceptions, turning apparent failures into successes, can be seen. Thus, "Mendelism", through the factorial hypothesis in the interpretation provided by the presence-and-absence hypothesis, explains the interaction of factors, that is, that factors are not only separate and underlying elements with individual and isolated effects, but they can also interact with each other and produce this way completely new characters (Bateson & Punnet 1905, Bateson 1909). A classic example that came to be widely known because it was the first time the (practical symbology of the) presence-and-absence theory was

---

analyzed mathematically this possibility in 1904, working from the assumption that the postulated allelomorphs had equal and additive effects, that there was complete dominance, and that both allelomorphs from a pair (calling "protogene" the A allelomorph or element, "allogene" the a allelomorph or element, "protogenic" the AA zygote, "allogenic" the aa pair and "heterogenic" the Aa zygote) were equally frequent (Pearson 1904). Yule pointed in 1906 that the complete dominance assumption was not adequately justified and proposed that the quantitative continuous fluctuation was caused by a great number of pairs of independent allelomorphs, each of which would have an insignificant influence on the measured character (Yule 1906).

From 1908 and 1909, H. Nilsson-Ehle (1909) in Sweden and E.M. East in the United States (East 1910, East & Hayes 1911) made significant contributions that enabled to understand and deal with characters that were not "alternative", big and *discontinuous*, but appeared continually. They observed a number of characters that showed an almost continuous series of grades and differed from one another in quantity rather than in quality. An example of it was the color of wheat. In order to explain the observed relation of 63 reds to 1 white, Nilsson-Ehle and East proposed that coloration was conditioned not by a unique pair or alleles, but by a set of pairs acting jointly in an additive way. The concept of "multiple factors" – or "polygenes" as were named later – was finally accepted and incorporated within the frame of the genetic theory developed by Morgan and collaborators, because now the different factors could be analyzed separately according to "Mendelian" principles.

[20] Bateson (1902, p. 116).

tested, is the case of combs in fowl. Each variety of fowl possesses a characteristic type of comb; there are some of the type called "rose"; others have a comb called "pea"; others have the single comb of the wild forms. The crossings of rose and pea comb varieties with single comb varieties show that both the rose and pea combs dominate over single combs. However, when pea and rose are crossed, a new and interesting result obtains. The first generation is uniform, but all the F_1 individuals exhibit a new form of comb, known as "walnut". When F_1 walnut shaped birds are crossed with each other, in the F_2 generation there appear not only walnut, rose and pea shaped combs, but also single combs, in the proportions 9:3:3:1 respectively. In these crossings, then, walnut comb in F_1 and single comb in F_2 appear as novelty (and since the latter represents a character that belongs to the wild form, its appearance was also characterized as "atavism"). In a dihybrid crossing a 9:3:3:1 numerical proportion is to be expected. The F_1 form, the walnut comb, is determined, according to Bateson, by two factors R and P; its formula as homozygote is $RRPP$. If only the R factor is present, without P, then the rose comb is generated (formula $RRpp$); the P factor without R determines the appearance of the pea comb ($rrPP$). If, finally, both R and P are absent, the single comb form is generated ($rrpp$). The difficulties in the analysis of combs in fowl subside, then, if it is assumed that the walnut comb is composed of the rose comb and the pea comb, and that the single comb appears when the factors for rose and pea combs are lacking. What was characteristic of this curious crossing was the fact that various factors that were completely independent from each other ("compound characters", in Bateson's words) contributed to the production of one character. This hypothesis, however, was presented with serious difficulties when trying to use it to explain dominant mutations, i.e., when trying to conceive of the way a new dominant factor is generated.

Also, when Bateson, Saunders and Punnett found in dihybrid crossings cases in which the numerical proportions in F_2 parted completely from the usual 9:3:3:1 proportion, they explained such proportions through what they denominated factor "coupling" and "repulsion". The central feature of such phenomena was constituted by the fact that combinations of factors, just as they were introduced by progenitors, appeared more frequently – though not exclusively – in F_2 than they used to. In case AB were crossed with ab, so that one of the progenitors had both dominant factors and the other both recessive ones, the hybrid F_1 ($AaBb$) forms gametes with the combination of factors from the progenitors in larger number than those that possess the combination of factors Ab and aB; if, adittionaly, – from the viewpoint of the presence-and-absence hypothesis – only dominant factors are considered as "true", we can speak of *coupling* between A and B. In case Ab crosses with aB, the F_1 hybrid ($AaBb$) identical from the outside to the former, will form again in larger number those gametes that posses the combination of factors of the progenitors – Ab and aB –; the combination AB will appear less frequently, in such a way that – from the presence-and-absence hypothesis – we should speak of a *"repulsion* of dominant factors" (or of a "false allelomorphism").[21]

4. Mendelism: A research programme in genetics

The view developed by Bateson and his collaborators can be rightfully considered the first research programme in genetics.

The concept of *scientific research programme* is introduced by Imre Lakatos (1968) inspired by Karl Popper's *metaphysical research programme*,[22] and developed afterwards by him and some of his collaborators (Lakatos 1970, 1971, 1974, Zahar 1973, Lakatos & Zahar 1976). Such developments, however, are not restricted to trying to characterize this metascientific concept with precision. They have tried to

---

[21] Since 1911, Bateson tried to explain the phenomena of coupling and repulsion through a segregation of pairs of factors that would take place during the first embryonic states of the plant, and a reproduction (*reduplication*) of certain types of gametes during their formation (*theory of reduplication*), that is, through cytological hypothesis.

[22] This term first appears in the *Postscript: Twenty Years After* that Popper writes to be published as appendix or volume added to the english version (published in 1959) of his *Logik der Forschung* (1934). This text was written mainly during the years 1951-1956, corrected and extended when Popper recovered his eyesight after surgery due to retina detachment, and barely modified after 1962. However, it was finally published during the years 1982-1983 in a three-volume edition prepared by W.W. Bartley III (Popper 1982, 1982a, 1983). On the Popperian concept of metaphysical research programme, see also Popper (1974, §§ 33, 37).

show that it is adequate for a better understanding of science and its history by means of applying it to several scientific domains, specially physics and chemistry.[23] Even though it has never been a privileged area, biology has not been completely foreign to the use of the concept of research programme in its analysis. So, *e.g.*, it is used by Michod (1981) to analyse the history of population genetics and, more closely related to the present work, by Meijer (1983), Van Balen (1986, 1987) and Martins (2002) to analyze the history of what is called "classical", "formal" or "Mendelian genetics". Although Lakatos' concept of research programme is also being used here, the analysis offered in this work differs from those provided by these authors, not only in the way the concept is understood, but also in the particular analysis of the history of genetics performed with it. Thus, Meijer (1983) thinks that Bateson's work can be described as a research programme *à la* Lakatos, though in a wider frame that includes the totality of the period 1900-1915, considering it as a *complete change of perspective* in heredity, but he does not make a systematic use of the notion – he does not identify explicitly the elements of a Lakatosian research programme in Bateson's work. Van Balen (1986, 1987), using a *modified concept* of research programme, in which Lakatos' *heuristics* (*positive* and *negative*) are replaced by Nickles' (1980, 1981) *constraints*,[24] presents Bateson's Mendelism as a research programme in which there is a theory of heredity and a saltationist conception of evolution that are inextricable, and he believes it is precisely this feature that distinguishes the "old (evolutionary) Mendelian programme" – Bateson's Mendelism – from the view developed by Morgan and his disciples, what he denominates "the new 'Mendelian genetics'". As has already been said, the problem of variation and heredity – what the discipline called "genetics" is concerned with – becomes independent from the problem of evolution, as well as from the problems of embryology, and acquires new identity and its own interests explicitly around 1905-1906 through Bateson's work, and not through the work done by

---

[23] In addition to the already mentioned texts by Lakatos (1970), Zahar (1973) and Lakatos & Zahar (1976), see, *e.g.*, Moulines (1989).

[24] On "heuristics," see below; Nickles introduces the term "constraint" to refer to any item of information, any "law", principle, rule or fact more or less established or accepted, that helps delimit a problem laying a condition for its solution.

Morgan and his collaborators, which was carried out from 1910 onwards.[25] Martins (2002), finally, thinks it is "more adequate to speak of the *Mendelian research programme* adopted by Bateson's group" rather than "[t]he Mendelian theory", because this theory "was highly malleable and subject to profound alterations", and tries to make it explicit through the characterization of a "defined experimental method, some basic concepts and a general theoretical framework" (p. 50), based on the analysis of Bateson's works from early in the twentieth century, such as Bateson (1901a), Bateson (1902) and Bateson & Saunders (1902). However, just like was pointed out in the case of Meijer, she does not make a systematic use of the Lakatosian concept of scientific research programme, that is, she does not identify explicitly in Bateson's work the elements that constitute a research programme. Moreover, as was already pointed out in this paper, the period of emergence of the research programme developed by Bateson and collaborators is situated between the years 1900 and 1905, and it can only be clearly identified as such programme from this latter date on and not in previous publications.

In what follows we will try to perform an analysis of the "Mendelism" developed by Bateson and collaborators by means of the Lakatosian concept of research programme. For that purpose, let us first indicate what this consists of.

For Lakatos, the unit of metatheoretical analysis is not an isolated hypothesis or theory (understood as a conjunction of hypothesis), but what he calls a *research programme*. Every research programme consists of a *hard core* – that vertebrates it and gives it unity – "tenaciously protected from refutation by a vast 'protective belt' of auxiliary hypothesis" (Lakatos 1978, p. 4). Programmes can also be characterized by the *heuristics* associated with the core, which consist in methodological rules of two sorts: "some tell us what paths of investigation to avoid (*negative heuristic*), and others what paths to pursue (*positive heuristic*)" (Lakatos 1978, p. 47). The programme's negative heuristic also forbids, *by a methodological decision*, the refutation of the core, "forbids us to direct the *modus tollens* at this 'hard core'" (Lakatos 1978, p. 47). Hence, we must "articulate or even invent 'auxiliary hypothesis',

---

[25] For a more detailed analysis of the relationship between Bateson's and Morgan's views, see Lorenzano (1998b, 2002a).

which form a *protective belt* around this core, and we must redirect the *modus tollens* to *these*" (Lakatos 1978, p. 48). The positive heuristics, on the other hand, "consists of a partially articulated set of suggestions or hints on how to change, develop the 'refutable variants' of the research-programme, on how to modify, sophisticate, the 'refutable' protective belt" (Lakatos 1978, p. 50).

Scientific research programmes ruled by a hard core are developed through changes in the protective belt of auxiliary hypothesis, thus providing a succession of different variants of the same programme. Lakatos offers also a typology of scientific research programmes, based on how "successful" they are:

> A research programme is said to be *progressing* as long as its theoretical growth anticipates its empirical growth, that is, as long as it keeps predicting novel facts with some success (*'progressive problemshift'*); it is *stagnating* if its theoretical growth lags behind its empirical growth, that is, as it gives only *post-hoc* explanations either of chance discoveries or of facts anticipated by, and discovered in, a rival programme (*'degenerating problemshift'*) (Lakatos 1978, p. 112).

From the constituents of a research programme I will take those that can be formalized or made precise with the aid of model-theoretic devices in a plausible way[26] and that will prove useful for an analysis of Bateson's Mendelism: the notion of irrefutable hard core distinguishable from a protective belt, built up with the aid of a positive heuristic, and the idea of progressiveness.

Let us now see how to apply these notions to the case of Bateson's Mendelism. We will start with the hard core. If Mendelism can be regarded as a research programme, we should be able to identify an irrefutable hard core, according to the negative heuristic, valid throughout the historical development of the programme and that guides the investigation, making possible through the positive heuristic the construction of a protective belt for such core by means of the articulation of more specific hypothesis.

---

[26] See Stegmüller (1973), Moulines (1979) and Balzer, Moulines & Sneed (1987) for a sharpening of some Lakatosian notions by means of the semantic or model-theoretic conception of theories known as "structuralist conception". For an introductory exposition of such metatheory, see Díez & Lorenzano (2002).

However, neither in the works of Bateson, nor in those of his collaborators can we find an explicit statement of what Lakatos calls "hard core". Are we to regard this situation, that is not exclusive to genetics in the biological sciences, as showing that there are no hard cores and ultimately no research programmes in this discipline? We think not. Rather, this core is present, though not explicitly articulated, unifying the programme and giving meaning to the practice of Mendelians, in such a way that if one wants to *understand* Bateson's view (i.e., Mendelism) and its development, one needs to *postulate* the existence of such core based on systematic reasons, thus making explicit what was only implicit. What would such a core consist in? Which would be its elements? This core[27] contains as basic elements the set of individuals (both parents and offspring), the set of traits or characters and the set of factors (present in the individuals by "allelomorphic" pairs of two sorts: that which denotes the presence of a factor and that which denotes its absence). In addition, through its articulation it establishes that for every parental pair that crosses and produces offspring, the probability distributions of factors in the offspring must coincide approximately with the relative frequencies of the characters observed in them, given certain relations between factors and characters.

This core, that the community of Mendelians accepts and uses throughout the development of the research programme, is highly schematic and general, having so little empirical content that turns out – in accordance with the negative heuristic of research programmes *à la* Lakatos – to be irrefutable.[28] Because, if – as happens here – the relative frequency of characters is determined empirically and the distribution of factors is postulated hypothetically, testing what the core states, namely, that the coefficients in the distribution of characters and in that of factors in the offspring are (approximately) equal, is a

---

[27] The exposition of such core is based on the following previously published works: Balzer & Dawe (1990), Balzer & Lorenzano (2000), Lorenzano (1995, 1998b, 2000a, 2002b).

[28] The irrefutable character of the core, however, is a matter of fact and not, as Lakatos would want it, the result of a methodological decision, unless we understand by that the decision to adopt such core, and not that to agree in not to direct the *modus tollens* at it.

paper and pencil task and does not involve any kind of empirical work. However, as happens with every core, despite being itself irrefutable, by connecting the different basic concepts, both the theoretical ones and the empirically accessible – in this case, the set of factors, probability distributions of factors in the offspring and the postulated relations between factors and characters, on one side, and the individuals, the set of characters and the relative frequencies of the characters observed in the offspring, on the other – it provides a conceptual framework within which testable, and eventually refutable, empirical hypothesis can be articulated. For this the positive heuristic is required.

This heuristic determines the ways to specify the core in order to obtain the particular testable hypothesis that constitutes the protective belt. Thus, the positive heuristic of the Batesonian research programme determines that in order to account for the distributions of parental characters in the offspring we must specify: a) the number of pairs of factors involved (one or more), b) the way factors are related to characters (complete or incomplete dominance, codominance or epistasis), and c) the way parental factors are distributed in the offspring (with combinations of equiprobable factors or not). Now, when this three kinds of specifications are performed, we obtain hypothesis at which we can direct the *modus tollens*.[29] These hypothesis possess all the additional information not contained in the core, and, for that very reason, have a more restricted domain of application.

Finally, it could be said that, due to the extension of the field of its applications, first to the so-called "discontinuous" characters and then also to those called "intermediate" or "continuous", both within the vegetal and the animal realm, in the period of 1905 to 1910 Mendelism was a progressive research programme.

---

[29] The correctness of this way of interpreting "Mendelism" seems to be supported by the way Thomas Hunt Morgan himself understood it, as can be inferred from the following claims from 1909: "In the modern interpretation of Mendelism, facts are transform into factors as a rapid rate. If one factor will not explain the facts, then two are invoked; if two proved insufficient, three will sometimes work out. The superior jugglery sometimes necessary to account for the results are often so excellently 'explained' because the explanation was invented to explain them and then, *presto*! explain the facts by the very factors that we invented to account for them." (Morgan 1909, p. 365).

5. Conclusion

Some of the conceptual and methodological changes that took place within the study of the problem of heredity during the first decade of the twentieth century and that led to the establishment of genetics as an autonomous discipline were exposed in this work. In particular, the changes associated with the works of William Bateson and his collaborators concerning the determination of the "problem of variation and heredity" as the central problem to be addressed, the articulation of "Mendelism" – with its "factorial hypothesis" – to address such problem, and its application, were exposed. The acceptance of this programme by the scientific community coincided with the establishment of genetics as a discipline. On the other hand, there has been in this work a systematic use of the notion crafted by Lakatos of scientific research programme with the purpose of identifying such programme – its irrefutable hard core, its protective belt, formed with the aid of the positive heuristic, and its progressiveness –, with the intention of contributing to a better understanding of (the history of) such discipline.

References

Bailey, L.H., 1903, "Some Recent Ideas on the Evolution of Plants", *Science* XVII (429), Friday, March 20: 441-454.

Balzer, W. & C.M. Dawe (1990), *Models for Genetics*, München: Institut für Philosophie, Logik und Wissenschaftstheorie.

Balzer, W. & P. Lorenzano (2000), "The Logical Structure of Classical Genetics", *Zeitschrift für allgemeine Wissenschaftstheorie* 31 (2): 243-266.

Balzer, W., Moulines, C.U. & J.D. Sneed (1987), *An Architectonic for Science. The Structuralist Program*, Dordrecht: Reidel.

Barnes, S. & S. Shapin (eds.) (1979), *Natural Order: Historical Studies of Scientific Culture*, Beverly Hills: Sage.

Bateson, B. (1928), *William Bateson, F.R.S., Naturalist. His Essays & Adresses together with a short account of his life*, Cambridge: Cambridge University Press.

Bateson, W. (1894), *Materials for the Study of Variation, treated with special regard to Discontinuity in the Origin of Species*, London: Macmillan and

Co.; "Preface and Introduction", reprinted in Punnett (1928), pp. 211-308.

Bateson, W. (1900a), "Hybridisation and Cross-Breeding as a Method of Scientific Investigation", *Journal of the Royal Horticultural Society* 24: 59-66; reprinted in Bateson (1928), pp. 161-170.

Bateson, W. (1900b), "Problems of Heredity as a subject for Horticultural Investigation", *Journal of the Royal Horticultural Society* 25: 54-61; reprinted in Bateson (1928), pp. 171-180.

Bateson, W. (1901a), "Experiments in Plant Hybridization", *Journal of the Royal Horticultural Society* 26: 1-3; reprinted in Punnett (1928), pp. 1-3.

Bateson, W. (1901b), "Heredity, Differentiation, and Other Conceptions of Biology: A Consideration of Professor Pearson's Paper 'On the Principle of Homotyposis'", *Proceedings of the Royal Society* 69: 193-205; reprinted in Punnett (1928), pp. 404-418.

Bateson, W. (1902), *Mendel's Principles of Heredity. A Defence*, Cambridge: Cambridge University Press.

Bateson, W. (1903), "On Mendelian Heredity of Three Characters Allelomorphic to Each Other", *Proceedings of the Cambridge Philosophical Society* 12: 153-154; reprinted in Punnett (1928), pp. 74-75.

Bateson, W. (1904), "Presidential Address to the Zoological Section, British Association", reprinted in Bateson (1928), pp. 233-259, and partly as "Heredity and Evolution", in *Popular Science Monthly*, New York, pp. 522-531.

Bateson, W. (1905a), "Letter to Adam Sedgwick from 18.4.1905", reprinted in Bateson (1928), p. 93.

Bateson, W. (1905b), "Letter to *Nature*", *Nature* 71: 390.

Bateson, W. (1906a), "The Progress of Genetic Research. An Inaugural Address to the Third Conference on Hybridisation and Plant-Breeding", *Reports of the Third International Conference on Genetics, Royal Horticultural Society*, pp. 90-97; reprinted in Punnett (1928), pp. 142-151.

Bateson, W. (1906b), "A Text Book of Genetics. Review of J. P. Lotsy's *Vorlesungen über Deszendenztheorien*, 1 Theil, Jena, 1906", *Nature* 74: 146-147; reprinted in Punnett (1928), pp. 442-445.

Bateson, W. (1907), "Facts Limiting the Theory of Heredity", *Science* 26: 649-662; reprinted in Punnett (1928), pp. 162-177.

Bateson, W. (1909), *Mendel's Principles of Heredity*, Cambridge: Cambridge University Press, 1st edition March 1909; 2nd edition (unmodified) August 1909; 3rd edition (extended) 1913; 4th edition (almost unmodified) 1930.

Bateson, W. & E.R. Saunders (1902), "The Facts of Heredity in the Light of Mendel's Discovery", *Experimental Studies in the Physiology of Heredity. Reports to the Evolution Committee of the Royal Society* I: 125-160; reprinted in Punnett (1928), pp. 29-68.

Bateson, W., Saunders, E.R., Punnett. R.C. & H. Kilby (1905), "Notes on the Progress of Mendelian Studies", *Reports to the Evolution Committee of the Royal Society* II: 119-131; reprinted in Punnett (1928), pp. 121-134.

Bateson, W. & R.C. Punnett (1905), "A Suggestion as to the Nature of the 'Walnut' Comb in Fowls", *Proceedings of the Cambridge Philosophical Society* 13: 165-168; reprinted in Punnett (1928), pp. 135-138.

Bateson, W., Saunders, E.R. & R.C. Punnett (1905), "Further Experiments on Inheritance in Sweet Peas and Stocks: Preliminary Account", *Proceedings of the Royal Society* B 77: 236-238; reprinted in Punnett (1928), pp. 139-141.

Bateson, W., Saunders, E.R. & R.C. Punnett (1906), "Experimental Studies in the Physiology of Heredity", *Reports to the Evolution Committee of the Royal Society* III: 2-11; reprinted in Punnett (1928), pp. 152-161.

Bateson, W., Saunders, E.R. & R.C. Punnett (1908), "Experimental Studies in the Physiology of Heredity", *Reports to the Evolution Committee of the Royal Society* IV: 2-5; reprinted in Punnett (1928), pp. 183-187.

Bateson, W. & R.C. Punnett (1911a), "On Gametic Series involving Reduplication of Certain Terms", *Journal of Genetics* I: 293-302; reprinted in Punnett (1928), pp. 206-214.

Bateson, W. & R.C. Punnett (1911b), "On the Interrelations of Genetic Factors", *Proceedings of the Royal Society* B 84: 3-8; reprinted in Punnett (1928), pp. 215-220.

Buck, R.C. & R.S. Cohen (eds.) (1971), *PSA 1970. Boston Studies in the Philosophy of Science*, Vol. 8, Dordrecht: Reidel

Carlson, E.A. (1966), *The Gene: a Critical History*, Philadelphia: Saunders.

Churchill, F.B. (1974), "William Johannsen and the Genotype Concept", *Journal of the History of Biology* 7: 5-30.

Cock, A.G. (1973), "William Bateson, Mendelism and Biometry", *Journal of the History of Biology* 6: 1-36.

Coleman, W. (1970), "Bateson and Chromosomes: Conservative Thought in Science", *Centaurus* 15: 228-314.

Coleman, W. (1970-1980), "Bateson, William", in Gillespie (1970-1980), pp. 505-506.

Correns, C. (1900a), "G. Mendels Regel über das Verhalten der Nachkommenschaft der Bastarde", *Berichte der Deutschen Botanischen Gesellschaft* 18: 158-168.

Correns, C. (1900b), "Gregor Mendel's 'Versuche über Pflanzen-Hybriden' und die Bestätigung ihrer Ergebnisse durch die neuesten Untersuchungen", *Botanische Zeitung* 58 (Supp.): 229-235.

Cuénot, L. (1902), "La loi de Mendel et l'hérédité de la pigmentation chez les souris", *Archives de Zoologie Expérimental et Générale*, 3 (Series, 10, Notes et Revue): 27-30.

Darden, L. (1977), "William Bateson and the Promise of Mendelism", *Journal of the History of Biology* 10: 87-106.

Delage, I. (1908), *L'hérédité et les grandes pròblemes de la biologie générale*, Paris: Schleicher frères & Cie., 2nd edition.

East, E.M. (1910), "A Mendelian Interpretation of Variation that is Apparently Continous", *American Naturalist* 44: 65-82.

East, E.M. & H.K. Hayes (1911), "Inheritance in Maize", *Connecticut Agricultural Station Bulletin* 167: 1-141.

Farrall, L.A. (1975), "Controversy and Conflict in Science: A Case Study – The English Biometric School and Mendel's Laws", *Social Studies of Science* 5: 269-301.

Froggatt, P. & N.C. Nevin (1971a), "Galton's 'Law of Ancestral Heredity': its Influence on the Early Development of Human Genetics", *History of Science* 10: 1-27.

Froggatt, P. & N.C. Nevin (1971b), "The 'Law of Ancestral Heredity' and the Mendelian-Ancestrian Controversy in England, 1889-1906", *Journal of Medical Genetics* 8: 1-36.

Gavroglu, K. Goudaroulis, Y. & P. Nicolacopoulos (eds.) (1989), *Imre Lakatos and Theories of Scientific Change*, Dordrecht: Kluwer.

Gillespie, C.C. (ed.) (1970-1980), *Dictionary of Scientific Biography*, New York: Charles Scribner's Sons.

Hurst, R. (1949), "The R.H.S. and the Birth of Genetics", *Journal of the Royal Horticultural Society* 74: 377-393.

Johannsen, W. (1909), *Elemente der exakten Erblichkeitslehre*, Jena: Gustav Fischer, 1st ed., 2nd ed. 1913, 3rd ed. 1926.

Johannsen, W. (1911), "The Genotype Conception of Heredity", *American Naturalist* 45: 129-159.

Johannsen, W. (1923), "Some Remarks About Units in Heredity", *Hereditas* 4: 133-141.

Kim, K.-M. (1994), *Explaining Scientific Consensus. The Case of Mendelian Genetics*, New York: The Guilford Press.

Lakatos, I. (1968), "Criticism and the Methodology of Scientific Research Programmes", *Proceedings of the Aristotelian Society* 69: 149-186.

Lakatos, I. (1970), "Falsification and the Methodology of Scientific Research Programmes", in Lakatos & Musgrave (1970), pp. 91-195; reprinted in Lakatos (1978), pp. 8-101.

Lakatos, I. (1971), "History of Science and Its Rational Reconstructions", in Buck & Cohen (1971), pp. 174-182; reprinted in Lakatos (1978), pp. 102-138.

Lakatos, I. (1974), "Science and Pseudoscience" (Radio Lecture broadcast by the Open University on 30 June 1973), in Vesey, G.

(ed.), *Philosophy in the Open*, Open University Press, 1974; reprinted as introduction to Lakatos (1978), pp. 1-7.

Lakatos, I. (1978), *The Methodology of Scientific Research Programmes: Philosophical Papers*, Volume 1, edited by Worrall, J. and G. Currie, Cambridge: Cambridge University Press, pp. 8-101.

Lakatos, I. & A. Musgrave (eds.) (1970), *Criticism and the Growth of Knowledge*, Cambridge: Cambridge University Press.

Lakatos, I. & E.G. Zahar (1976), "Why Did Copernicus's Programme Supersed Ptolemy's?", in Westman (1976), pp. 354-383.

Lorenzano, P. (1995), *Geschichte und Struktur der klassischen Genetik*, Frankfurt/Main: Peter Lang.

Lorenzano, P. (1997), "Hacia una nueva interpretación de la obra de Mendel", in Ahumada, J. & P. Morey (eds.), *Selección de trabajos de las VII Jornadas de Epistemología e Historia de la Ciencia*, Córdoba: Facultad de Filosofía y Humanidades, Universidad Nacional de Córdoba, pp. 220-231.

Lorenzano, P. (1998[a]), "Acerca del 'redescubrimiento' de Mendel por Hugo de Vries", *Epistemología e Historia de la Ciencia* 4: 219-229.

Lorenzano, P. (1998b), "Hacia una reconstrucción estructural de la genética clásica y de sus relaciones con el mendelismo", *Episteme* 3 (5): 89-117.

Lorenzano, P. (1999), "Carl Correns y el 'redescubrimiento' de Mendel", *Epistemología e Historia de la Ciencia* 5: 265-272.

Lorenzano, P. (2000a), "Classical Genetics and the Theory-Net of Genetics", in Balzer, W. & C.U. Moulines (eds.) (2000), *Structuralist Knowledge Representation: Paradigmatic Examples*, Amsterdam: Rodopi, pp. 251-284.

Lorenzano, P. (2000b), "Erich Tschermak: supuesto 'redescubridor' de Mendel", *Epistemología e Historia de la Ciencia* 6: 251-258.

Lorenzano, P. (2002a), "Leyes fundamentales, refinamientos y especializaciones: del 'mendelismo' a la 'teoría del gen'", in Lorenzano, P. & F. Tula Molina (eds.) (2002), *Filosofía e Historia de la Ciencia en el Cono Sur*, Bernal: Universidad Nacional de Quilmes, pp. 379-396.

Lorenzano, P. (2002b), "La teoría del gen y la red teórica de la genética", in Díez, J.A. & P. Lorenzano (eds.) (2002), *Desarrollos actuales de la metateoría estructuralista: problemas y discusiones*, Bernal: Universidad Nacional de Quilmes/Universidad Autónoma de Zacatecas /Universidad Rovira i Virgili, pp. 285-330.

MacKenzie, D. (1978), "Statistical Theory and Social Interests: A Case Study", *Social Studies of Science* 8: 35-83.

MacKenzie, D. (1979), "Karl Pearson and the Professional Middle Class", *Annals of Science* 36: 125-143.

MacKenzie, D. & S.B. Barnes (1975), "Biometriker versus Mendelianer. Eine Kontroverse und ihre Erklärung", *Kölner Zeitschrift für Soziologie und Sozialpsychologie*, Sonderheft 13: 165-196.

MacKenzie, D. & S.B. Barnes (1979), "Scientific Judgment: The Biometry–Mendelism Controversy", in Barnes & Shapin (1979), pp. 191-210.

Marrais, R. de (1974), "The Double-Edged Effect of Sir Francis Galton: A search for the Motives in the Biometrician-Mendelian Debate", *Journal of the History of Biology* 7: 141-174.

Martins, L.A.-C.P. (2002), "Bateson e o programa de pesquisa mendeliano", *Episteme* 14: 27-55.

Masters, M. (1900), "Societies: Royal Horticultural Lecture", *Gardener's Chronicle* 27: 3.

Meijer, O.G. (1983), "The Essence of Mendel's Discovery", in Orel, V. & A. Matalová (eds.), *Gregor Mendel and the Foundation of Genetics*, Brno: The Mendelianum of the Moravian Museum in Brno, 1983, pp. 123-178.

Mendel, G. (1865), "Versuche über Pflanzen-Hybriden", *Verhandlungen des Naturforschenden Vereins zu Brünn* 4: 3-57; reprinted in *Ostwalds Klassikern der exakten Wissenschaften,* Nr. 6, Braunschweig: Friedr. Vieweg & Sohn, 1970.

Michod, R.E. (1981), "Positive Heuristics in Evolutionary Biology", *British Journal for the Philosophy of Science* 32: 1-36.

Morgan, T.H. (1909), "What are Factors in Mendelian Inheritance?", *American Breeders' Association Report* 6: 365-368.

Morgan, T.H. (1913), *Heredity and Sex*, New York: Columbia University Press.

Morgan, T.H. (1916), *A Critique of the Theory of Evolution*, Princeton, Princeton University Press.

Moulies, C.U. (1979), "Theory-Nets and and the Evolution of Theories: The Example of Newtonian Mechanics", *Synthese* 4: 417-439.

Moulies, C.U. (1989), "The Emergence of a Research Programme in Classical Thermodynamics", in Gavroglu, Goudaroulis & Nicolacopoulos (1989), pp. 111-121.

Nägeli, C.v. (1884), *Mechanisch-physiologische Theorie der Abstammungslehre*, München u. Leipzig: R. Oldenburg.

Nickles, T. (1980), "Can Scientific Constraints Be Violated Rationally?", in Nickles (1980), pp. 285-315.

Nickles, T. (ed.) (1980), *Scientific Discovery, Logic and Rationality*, Dordrecht: Reidel.

Nickles, T. (1981), "What Is a Problem That We May Solve It?", *Synthese* 47: 85-118.

Nilsson-Ehle, H. (1909), "Kreuzungsuntersuchungen an Hafer und Weizen", *Lunds Universitets Årsskrift* 52.

Norton, B.J. (1973), "The Biometric Defense of Darwinism", *Journal of the History of Biology* 6: 283-316.

Norton, B.J. (1975a), "Biology and Philosophy: The Methodological Foundations of Biometry", *Journal of the History of Biology* 8: 85-93.

Norton, B.J. (1975b), "Metaphysics and Population Genetics: Karl Pearson and the Background to Fisher's Multi-factorial Theory of Inheritance", *Annals of Science* 32: 537-553.

Norton, B.J. (1978), "Karl Pearson and Statistics: The Social Origins of Scientific Innovation", *Social Studies of Science* 8: 3-34.

Olby, R. (1987), "William Bateson's Introduction of Mendelism to England: A Reassessment", *British Journal for the History of Science* 20: 399-420.

Olby, R. (1988), "The Dimensions of Scientific Controversy: The Biometric-Mendelian Debate", *British Journal for the History of Science* 22: 299-320.

Olby, R. (2002), "Mendelism: From Hybrids and Trade to a Science", in Lorenzano, P. & F. Tula Molina (eds.), *Filosofía e Historia de la*

Ciencia en el Cono Sur, Bernal: Universidad Nacional de Quilmes, pp. 21-39.

Pearson, K. (1904), "Mathematical Contributions to the Theory of Evolution. XII: On a Generalised Theory of Alternative Inheritance, with Special Reference to Mendel's Laws", *Philosophical Transactions of the Royal Society of London* A 203: 53-86.

Popper, K. (1934), *Logik der Forschung*, Wien: Julius Springer Verlag.

Popper, K. (1974), "Intellectual Autobiography", in Schilpp, P.A. (ed.), *The Philosophy of Karl Popper*, La Salle, Ill.: Open Court, pp. 3-181.

Popper, K. (1982), *The Open Universe. An Argument for Indeterminism*, Vol. II of the *Poscript to The Logic of Scientific Discovery*, London: Hutchinson & Co.

Popper, K. (1982a), *Quantum Theory and the Schism of Physics*, Vol. III of the *Poscript to The Logic of Scientific Discovery*, London: Hutchinson & Co.

Popper, K. (1983), *Realism and the Aim of Science*. Vol. I of the *Poscript to The Logic of Scientific Discovery*, London: Hutchinson & Co.

Provine, W.B. (1971), *The Origins of Theoretical Population Genetics*, Chicago: The University of Chicago Press.

Punnett, R.C. (1905), *Mendelism*, Cambridge: Macmillan and Co, 1st ed., 2nd ed., 1907, 3rd ed. 1911, 4th ed. 1912, 5th ed. 1919, 6th ed. 1922, 7th ed. 1927.

Punnett, R.C. (1950), "Early Days of Genetics", *Heredity* 4: 1-10.

Punnett, R.C. (1952), "William Bateson and Mendel's Principles of Heredity", *Notes and Records of the Royal Society of London* 9: 336-347.

Punnett, R.C. (ed.) (1928), *Scientific Papers of William Bateson*, Cambridge: Cambridge University Press.

Roll-Hansen, N. (1978), "The Genotype Theory of Wlhelm Johannsen and its Relation to Plant Breeding and the Study of Evolution", *Centaurus* 22: 201-235.

Roll-Hansen, N. (1980), "The Controversy Between Biometricians and Mendelians: A Test Case for the Sociology of Scientific Knowledge", *Social Science Information* 3: 501-517.

Roll-Hansen, N. (1989), "The Crucial Experiment of Wilhelm Johannsen", *Biology and Philosophy* 4: 303-329.

Stegmüller, W. (1973), *Theorienstrukturen und Theoriendynamik*, Berlin: Springer.

Swinburne, R.G. (1965), "Galton's Law-Formulation and Development", *Annals of Science* 21: 15-31.

Tschermak, E. (1900), "Über künstliche Kreuzung bei Pisum sativum", *Berichte der Deutschen Botanischen Gesellschaft* 18: 232-239.

Van Balen, G. (1986), "The Influence of Johannsen's Discoveries on the Constraint-Structure of the Mendelian Research Program. An Example of Conceptual Problem Solving in Evolutionary Theory", *Studies in History and Philosophy of Science* 17: 175-204.

Van Balen, G. (1987), "Conceptual Tensions Between Theory and Program: The Chromosome Theory and the Mendelian Research Program", *Biology and Philosophy* 2: 435-461.

Vries, H. de (1900a), "Sur la loi de disjonction des hybrides", *Comptes Rendus de l'Académie des Sciences* 130: 845-847.

Vries, H. de (1900b), "Das Spaltungsgesetz der Bastarde (Vorläufige Mittheilung)", *Berichte der Deutschen Botanischen Gesellschaft* 18: 83-90.

Vries, H. de (1900c), "Sur les unités des caractéres spécifiques et leur application a l'étude des hybrides", *Revue générale de Botanique* 12: 257-271.

Wanscher, J.H. (1975), "The History of Wilhelm Johannsen's Genetical Terms and Concepts from the Period 1903 to 1926", *Centaurus* 19: 125-147.

Westman, R. (ed.) (1976), *The Copernican Achievement*, Los Angeles: University of California Press.

Yule, G.U. (1906), "On the Theory of Inheritance of Quantitative Compound Characters on the Basis of Mendel's Law–A Preliminary Note", *International Conference on Hybridisation and Plant Breeding. Report of the Third International Conference on Genetics*, London: Royal Horticultural Society, pp. 140-142.

Zahar, E. (1973), "Why Did Einstein's Programme Supersede Lorentz's? (I) & (II)", *British Journal for the Philosophy of Science* 24: 95-123, 223-262.

On the Origin of «That Thing You Call "Species"»·

Santiago Ginnobili

University of Buenos Aires (UBA)/
National University of Quilmes (UNQ)/
National Scientific and Technical Research Council (CONICET)

1. Introduction

This paper analyzes the relationship between the theoretical term "species" and the theory of natural selection as Darwin conceived it. For this purpose, use is made of the distinction provided by structuralism between terms introduced in a theory and theoretical terms previously available. The thesis of this paper is that "species" is not a term that depends semantically on the theory of natural selection (in a sense that will be clarified later), but one previously available instead. In structuralist language: "species" is *Natural Selection-non theoretical*. This analysis intends also to illustrate the benefits of a moderate ver-

---

· I am thankfull to Martín Ahualli and Rodrigo Moro for their comments to a previous version of this work. Also to all the members of the research group on scientific realism directed by Rodolfo Gaeta for their comments and criticisms. Finally I must give very special thanks to Pablo Lorenzano. Not only for having made valuable comments after a careful reading of the text, but also for introducing me to structuralist ideas, on which this work is mostly inspired.

sion of holism regarding the meaning of theoretical terms as an alternative to radical holism.

In *The Disorder of Things* Dupré advocates, among other things, a pluralist answer to the question concerning the ontological status of species. One of the arguments he offers is the following (Dupré 1993, pp. 38-39):

a) "Species" is a theoretical term.
b) Theoretical terms are to be understood by means of the theoretical context in which they occur.
c) Species are treated as individual objects in the central parts of evolutionary biology, and as classes in ecology.

Therefore,

d) we are driven towards a pluralist view on the ontological status of species. In some contexts they are treated as classes, in others as individuals.

This paper does not intend either to analyze this argument at full length, or to criticize pluralism. I will only focus on the assumptions underlying premise *b*, which asserts that theoretical terms cannot be understood without paying attention to the theoretical contexts in which they occur. I assume I am not doing violence to Dupré's thought by claiming that this premise presupposes the thesis that theoretical terms acquire at least part of their meaning from the relationships they hold with the rest of the terms in the theories in which they appear. I do not intend to criticize holism regarding the meaning of theoretical terms either; however, I find it more fruitful to restrict this generalized version of holism to a more localized version. The claim in premise *b* does not consider the possibility that theoretical terms might not acquire their meaning from *all* the theories in which they appear. Specially, it does not take into account the fact that there are two ways in which one and the same theoretical term may appear in a theory: it may be a term introduced by that theory and whose application presupposes the theory, or it may be a theoretical term previously available and that can be applied without any reference to the theory. In *Philosophy of Natural Science*, Hempel calls the former "theoretical terms proper", and the latter "pretheoretical or antecedently available terms" (Hempel 1966, pp. 74-75). This distinction is relative to a theory. A term that is properly theoretical in a theory may be pre-theoretic in another and vice versa. If this distinction is correct, then it is not true that the meaning of theoretical terms depends on all the theories in which they might appear. It would depend only on the theories for which they are properly theoretical.

Structuralism offers a distinction between theoretic for a given theory T and non theoretic for a given theory T, which is the one I am going to use, albeit in an informal way, because I think it elucidates Hempel's intuition.

2. *T*-theoreticity

Structuralism rejects the traditional theoretical/observational distinction. This distinction encloses two different ones: theoretical/non-theoretical, and observational/non-observational. Of these, structuralism only preserves the first, though, as we have seen, relative to a given theory.[1] As Moulines explains in *Pluralidad y recursión* (1991), opposing both *operationalism* – which equates the meaning of a theoretical term with the physical processes that can be associated with it – and *radical semantical holism* – which claims that the meaning of a theoretical term is determined by every theory in which the term appears (the kind of holism Dupré seems to uphold) –, a *moderate holism* holds. According to it, there are terms – the T-theoretical ones – that depend semantically on a given theory T, and others – the T-non-theoretical – than do not depend semantically on T, but on another theory and may be used to contrast T (Moulines 1991 ch. II.3). Saying that a term is semantically dependent on a given theory T means that in order to determine the concept expressed by such term it is always necessary to assume the validity of T's laws, in which case the term is T-theoretical. A T-non-theoretical term is such that it is not always necessary to assume T's laws for the determination of the concept expressed by it. The determination of a concept, in the case of qualitative ones, consists of determining whether it applies to a particular given object, and in the case of quantitative ones, it consists of deter-

---

[1] The reasons for the rejection of the theoretical/observational distinction as inadecuate for the task of reconstructing scientific theories and their empirical basis can be found in Balzer, Moulines & Sneed (1987, p. 48).

mining the value of the magnitude for the object (Díez & Moulines 1997, pp. 354-356).

3. Restating the question

It was already hard to think that the problem of the meaning of "species" might have a general solution for all branches of biology, given it presupposes a unity of the discipline that is difficult to sustain. In this framework, it does not make any sense either presenting the issue as relative to evolutionary biology in general. To elucidate the meaning of the term "species" it is necessary to look at the role played by the term in each theory. It might occur that as a result of the reconstruction of all the theories that constitute evolutionary biology we reach a single concept of *species*, but this is a result of meta-theoretic investigation. The unity of evolutionary biology must not be presupposed.

Therefore, the problem of the meaning of "species" must be presented in relation to particular theories. Though this way of dealing with the problem requires having systematized such theories in some fashion, I will now try to apply this analytical framework to Darwin's theory of natural selection in an informal manner intending to show how a term might appear in a theory without being theoretical for that theory.

4. "That thing you call 'species'"

The most extended opinion is that Darwin was a nominalist concerning the concept of *species*.[2] Mayr, for instance, holds that:

> [...] his characterization of the species is now [he refers to the time of the *Origin*] a mixture of typological and nominalist definitions. (Mayr 1991, p. 30)

These opinions find strong support in Darwin's own words. In the *Origin* he makes several statements like the following:

---

[2] I am speaking of Darwin's opinions in the *Origin*. Darwin changed several times his opinions about the concept of species until he reached his conceptions of the *Origin*.

> In short, we shall have to treat species in the same manner as those naturalist treat genera, who admit that genera are merely artificial combinations made for convenience (Darwin 1859, p. 485).

Statements like this one seem to support the idea that in the context of natural selection "species" does not appear at all, but I find this presumption too hasty. What Darwin is really opposing to is the search for essences founding the differences between species and the search for an essence of the term "species" itself, and consequently, the possibility of defining the species and the term "species". This rejection is really a rejection of fixist and creationist ideas. But the fact that a term cannot be defined does not mean it cannot be theoretical for a given theory. On the contrary, their openness, that is, their lack of definition, is for some one important feature of the most fruitful theoretical terms (Hempel 1952, pp. 28-29). Nevertheless, given that Darwin did not consider the fact that the term "species" lacked a definition to be an objection to his theory of natural selection, and that conventionalist elements played some role in the determination of its extension, it seems reasonable to assume that he did not think the term "species" played an important role in his theory. Especially considering he not only did not regard them as objections, but rather as positive consequences of his theory:

> Systematists will be able to pursue their labours as at present; but they will not be incessantly haunted by the shadowy doubt whether this or that form be in essence a species. This I feel sure, and I speak after experience, will be no slight relief (Darwin 1859, p. 484).

The claim that the term "species" does not play a very important role might seem quite difficult to sustain. The term "species" appears hundreds of times in the *Origin*, it even appears in its title. However, whatever conception of theory is presupposed, it seems fairly obvious that we have to distinguish between the theory of natural selection and the book in which this theory appears along with others (according to Mayr there are at least five theories that are held in the *Origin* (Mayr 1991, pp. 35-37). Besides, it would not be the first time the title of a book is misleading about its contents.

On the other hand, it is interesting to bring forward the proposal Beatty takes from Sulloway (Sulloway 1979) concerning why Darwin uses the concept of *species* and how he does so (Beatty 1985). Darwin's decision to use the concept of *species* would have been guided by tactical considerations. Among these would be the decision to employ the same concept that naturalists of his time employed in order to communicate with them in their own language. According to Beatty, Darwin used the term "species" in the same way his contemporaries did, but without accepting the definition they ascribed to it, which was incompatible with his evolutionary views. Darwin, then, would have used "species" to refer to the same things in the world his contemporary naturalists referred to, but believed these things did not satisfy the given definition of "species". Accepting its reference while rejecting its definition allowed him to communicate with naturalists and also provided him some space for disagreement. Not only did Darwin not accept the commonly accepted definitions of "species", but also held that the term was indefinable, as we saw earlier. Species could not be clearly distinguished from varieties, fact that is intelligible in the light of diversifying evolution, in which varieties are incipient species. The following quote from Darwin can be used in support of Beatty's interpretation:

> At the end of this chapter, it will be seen that according to the views, which we have to discuss in this volume, it is no wonder that there should be difficulty in defining the difference between a species & a variety; there being no essential, only an arbitrary difference. In the following pages I mean by species, those collections of individuals, which have commonly been so designated by naturalists (in Stauffer 1975, p. 98).

If this interpretation is correct it would help explain why the term "species" appears so many times in the *Origin* without having to admit that the concept of species plays a fundamental role in the theory of natural selection.

But let us be more systematic. Let us leave Darwin's opinions aside and ask ourselves: is "species" a NS-theoretical (*Natural Selection*-theoretical) term? We should remember that a term is T-theoretical if we must always suppose the applicability of T's fundamental laws in order to determine the term's extension. So, are the fundamental laws of natural selection always supposed whenever the extension of "species" is determined? Given the multiplicity of concepts expressed by "species", I will focus on some of the most salient ones to answer this question. I repeat: for an exhaustive treatment of this issue we need to have the theory of natural selection reconstructed, but I think that treating it before this reconstruction might still be productive. We will consider then the typical informal presentation of natural selection.

Let us start with the *morphological concept of species*. According to this concept, a species is a collection of individuals who posses similar morphological features. We can include within this concept the typological concept of *species*, in which the morphological properties shared by all the individuals must be essential to the species in question. It is clear that the determination of whether an individual belongs to a species is in this case independent of natural selection. Being included or not in a species depends on the presence or absence of certain features in a certain degree, and I cannot think how confirming that an individual possesses those features might depend on natural selection. This does not mean that we cannot choose a concept of *morphospecies* compatible with natural selection rejecting, for instance, any atavistic essentialist feature of that concept – the fenetist concept of *species*, for example. But a concept's being compatible with a theory is not enough to make it theoretical in that theory. To make it so it should be impossible to determine the extension of the concept without resorting to the theory's laws.

As for the *biological concept of species*, whether we refer to a concept in which a collection of individuals is a species if each of the individuals can have a fertile descendance with any of the others, or whether we refer to a more modern one that applies to populations and not to individuals, in which a species is constituted by populations connected by genic flux and reproductively isolated from other populations, it is quite clear that such concept is not NS-theoretical either. In both cases we find criteria whose application is clearly independent of natural selection. Although this concept has major importance for evolutionists, it is not necessary to suppose natural selection to determine whether two individuals can have fertile offspring, or whether two populations are connected by genic flux.

Finally, let us consider the *evolutionary concept of species*. This concept is more interesting because it was apparently the most attractive one for Darwin. It is held in various places that the classificatory system should really reflect the genealogical relationships amongst different individuals:

> According to my opinion, (which I give every one leave to hoot at, like I should have, six years since, hooted at them, for holding like views) classification consists in grouping beings according to their actual relationship, ie their consanguinity, or descent from common stocks (Burkhardt 1996, p. 76).

According to this view, species are segments of the filogenetic tree with certain characteristics. Jean Gayon argues that in the *Origin* Darwin favored a concept of species of this sort (Gayon 1996). Near the end of the *Origin* Darwin describes what the task of systematists would be like once his evolutionary views were accepted:

> Systematists will have only to decide (not that this will be easy) whether any form be sufficiently constant and distinct from other forms, to be capable of definition; and if definable, whether the differences be sufficiently important to deserve a specific name (Darwin 1859, p. 484).

And then he claims that the present degree of difference should be measured more carefully, because:

> It is quite possible that forms now generally acknowledged to be merely varieties may hereafter be thought worthy of specific names (Darwin 1859, p. 485).

This quote is interpreted by Mayr as showing that Darwin held a concept of *species* that was halfway between nominalistic and typological (Mayr 1991, p. 30). I agree with Gayon that the criteria provided here by Darwin are such that they allow the systematic to infer genealogical links. The assumed concept of *species* would then be the evolutionary one, rather than the typological one. But Gayon's interpretation goes beyond this claim. The criteria mentioned in the quote from Darwin would not count as empirical rules for the recognition of species, but as theoretical claims that would furnish "species" with meaning in relation to the principles of natural selection. Morphological differences should be then considered in the light of natural selection. There would not be, according to Gayon, any definition of 'species' independent of the theory of natural selection. It would seem then that this concept were indeed *NS*-theoretical. But I think Gayon exaggerates when he infers a semantical dependence from that quote form Darwin. Darwin is telling there how the different branches of natural history will be affected once his views are accepted. I agree that Darwin held that what the systematics understood by "species" could no longer be the same. But there is a great distance from that to the assertion of a meaning dependence. The systematics might search for a concept of *species* compatible with Darwin's theory and yet with an independent meaning. Besides, in any case, I think that the fact that varieties are incipient species, something that is supposed by this whole discussion, does not depend on natural selection. I believe Mayr is right to distinguish the different theories supported by Darwin in the *Origin*. The ones supposed in this case are the theory of the common origin – that all organisms descend from a common ancestor –, without which it would not be possible to construct a single filogenetic tree, and the theory of diversifying speciation – that species diversify into daughter species, those that are at first varieties of a species can end up being daughter species. None of these two theories follows necessarily from natural selection.

Notwithstanding what Gayon affirms, let us examine the question of whether the evolutionary concept of *species* is natural selection-theoretical. According to this concept, I repeat, species are segments with certain features in a genealogical tree, that is, a species would be the segment contained between a speciation event – the birth of a species – and an extinction or speciation of the species event. Once the filogenetic tree is complete, there is no appeal to the laws of natural selection to determine the extension of "species". We only have to distinguish in the filogenetic tree those branches that posses distinctive features. We could ask then whether it is necessary to resort to the theory in the construction of the filogenetic tree itself. Again, we must give a negative answer. Although the acceptance of natural selection might influence the manner in which the filogenetic tree is framed, it is constructed out of morphological similarities among individuals (as Darwin recommends in the fragments quoted above), fossil record, and finally – though not in Darwin's time, of course – data provided by molecular biology. Therefore, the evolutionary concept of *species* is not *NS*-theoretical either.

5. "Species" as a *NS*-non-theoretical term

As we have seen, there are reasons to believe that "species" is not *NS*-theoretical. We should inquire now if the term "species" appears in the theory as a *NS*-non-theoretical term. I do not intend to give this issue full treatment here, but only to suggest two possible ways in which this might occur.

The first one takes place in a strict Darwinian frame. In chapter VI of the *Origin* Darwin presents and tries to answer different objections to the theory of natural selection. The first of these is the following:

> why, if species have descended from other species by insensibly fine gradations, do we not everywhere see innumerable transitional forms? Why is not all nature in confusion instead of the species being, as we see them, well defined? (Darwin 1859, p. 171).

Though natural selection was not for Darwin the only mechanism of evolution, it certainly was the main one. This means that natural selection was mainly responsible for explaining the forms in which living organisms presented themselves. But it turns out that in nature living organisms present themselves in discrete groups. Natural selection should explain that. In the empirical basis used for contrasting natural selection we find discrete groups of organisms, that is, we find species. This is one way in which "species" might appear in natural selection as a *NS*-non-theoretical term.

The second way is more or less foreign to Darwin's thought. For him, the unit of selection was mainly the individual (although in some places in the *Origin* he suggests that groups of individuals might also play that role). However, there is no impediment for natural selection to work on other levels. There are three requisites for a given entity to evolve by means of natural selection: a) it must vary in its fenotypical features, b) these features must be inheritable, and c) these fenotypical variations must correspond to differences in survival and reproduction (Sober & Wilson 1998, p. 83). If species satisfy these requisites nothings prevents them from evolving by natural selection. So, another possibility for "species" to fit into the context of natural selection as a *NS*-non-theoretical term is as the unit for selection.

6. Conclusions

If my argument is correct, then there is sound reason to believe that the meaning of the term "species" does not depend on natural selection. People often speak of the problem of species without being explicit as to what that problem is. Just as I find it useful to relatate the question of theoreticity to a given theory, I find it useful to do the same with problems. A problem is only such in the light of some theory. If by "the problem of species" we mean the inability to find a criterion that distinguishes it from inferior and superior taxa, we have already seen this is not a problem for natural selection (at least it was not for Darwin). What did and does constitute a problem is the absence of transitional varieties, that is, the fact that organisms come in species, and remains to be a problem also the question of whether species can be the unit for selection.

In case I have not provided sufficient argument for the central thesis of this work, I hope at least to have shown the relevance of the distinction between theoretical and non-theoretical terms in relation to a given theory for the question of the meaning of the term "species".

References

Balzer, W., Moulines, C.U. & J.D. Sneed (1987), *An Architectonic for Science. The Structuralist Program*, Dordrecht: Reidel.

Beatty, J. (1985), "Speaking of Species: Darwin's Strategy", in Ereshefsky, M. (ed.), *The Units of Evolution: Essays on the Nature of Species*, Cambridge, MA: The MIT Press, 1992, pp. 227-246.

Burkhardt, F. (ed.) (1996), *Charles Darwin's Letters. A Selection 1825-1859*, Cambridge: Cambridge University Press.

Darwin, C. (1859), *On the Origin of Species*, London: John Murray. (Facsimile edition, E. Mayr (ed.), Cambridge, MA: Harvard University Press, 1964.)

Díez, J.A. & C.U. Moulines (1997), *Fundamentos de Filosofía de la Ciencia*, Barcelona: Ariel.

Dupré, J. (1993), *The Disorder of Things*, Cambridge, MA: Harvard University Press.

Gayon, J. (1996), "The Individuality of the Species: A Darwinian Theory? – from Buffon to Ghiselin, and Back to Darwin", *Biology and Philosophy* 11: 215-244.

Hempel, C.G. (1952), *Fundamentals of Concept Formation in Empirical Science*, Chicago: University of Chicago Press.

Hempel, C. G. (1966), *Philosophy of Natural Science*, New Jersey: Prentice-Hall.

Mayr, E. (1982), *The Growth of Biological Thought*, Cambridge, MA: Harvard University Press.

Mayr, E. (1991), *One Long Argument*, Cambridge, MA: Harvard University Press.

Moulines, C.U. (1991), *Pluralidad y Recursión*, Madrid: Alianza.

Sober, E. & O.S. Wilson (1998), *Unto Others. The Evolution and Psychology of Unselfish Behavior*, Cambridge, MA: Harvard University Press.

Stauffer, R. (ed.) (1975), *Charles Darwin's Natural Selection: Being the Second Part of his Big Species Book written From 1856 to 1858*, London: Cambridge University Press.

Sulloway, F. (1979), "Geographic Isolation in Darwin's Thinking", *Studies in the History of Biology* 3: 23-65.

Alternatives in the Status of Darwinian Theory in Relation to the Epistemological Approach

Gladys Martínez

Susana La Rocca

Faculty of Humanities, ICEM GROUP
National University of Mar del Plata (UNMdP)

1. Introduction

The evolutionary theory in the Darwinian approach constitutes, no doubt, an important axis in the configuration of the biological disciplines. The epistemological incidence of this theory becomes specially apparent in its effective capacity to define the type of matters or questionings, the methodology, the explanatory model, etc. which will distinguish the biological sciences from their appearance, differentiating them from the physicalist prevailing paradigm of the 19th century.

However, not only the acceptance of all its thesis, but its prestige in the scientific community suffered from significant changes which allow us to identify, after its successful initial reception, the so-called "eclipse of the Darwinian theory", which took place at the end of the 19th century, and is followed by an application of restoration known as the "neo-Darwinism" in the 20th century. From then on, as it is sufficiently known, the Darwinian evolutionary theory does not only constitute the theoretical framework of most of the studies of Biology, but it also goes beyond the limits of the discipline achieving a paradigmatic character in relevant fields of present-day science. We take into account that the process by which a theory loses or gains an important status involves its complexity, epistemological interesting matters.

We furthermore suppose that an analysis of this character requires the contribution of different perspectives or epistemological patterns since it is not enough to cross-reference empirical and /or experimental matters but it requires taking into account the transformations produced in the structure of the practice and the scientific community such as the historical requirements which inform us on the situation of Biology in the stage we are interested in. The confluence of these dimensions collaborate in the understanding of the scientific activity allowing us to have a better approach to the real process, which is a matter of indubitable interest to the epistemological perspective. The main aim of this work is to evaluate the epistemological implications involved in the crisis and the later restoration of the Darwinian theory which took place at the end of the XIX century and the first half of the XX Century upon the basis of the transformations which are produced in the biological disciplines from the first Darwinian proposal and the synthesis which shapes its restoration.

2. Development

When in 1859 Darwin unveils his theory, he manages to move[1] the community of his time, which argues about his proposal in the most distinguished scientific fields. Nevertheless, it is important to take into account that such a repercussion was not equivalent to the acceptance of all the theses of the theory by the scientific community. On the contrary, we can establish a difference in its contributions from the epistemological point of view. In fact, it is possible to recognize a core of matters which we will name "*minimum* Darwinian proposal" which not only has an impact, but is adopted by researchers of biological processes. Correlatively, we can identify a "conception of maximum" which comprises the above-mentioned items, a set of theses

---

[1] In November of 1859 the *Origin of the Species* is published, and 1250 copies, 15 shillings each were sold out on the day of its publication.

which, at the beginning, are unacceptable to the paradigmatic conception of the world and of life at that moment (Kitcher 1993). A brief analysis of these two aspects of the theory proves to be of interest for these considerations.

1. In the framework of the *minimum* Darwinian proposal, relevant novelties which have a deep influence in the biological research work of that time are included and are specially related to:

- The type of queries, which are proposed, and which give a new character to the scientific practice in the field of biological studies. In the specific case of the evolutionary theory, the questions which arise alter the science-theology relationship accepted at the time, posing the question on the origin of species, since it asks not only about the appearance of the different bio-geographical distributions (Darwin 1872, p. 373, p. 469), but also the concept of standardization (Darwin 1872, p. 433) and analogy (Darwin 1872, pp. 424-426). From this framework, the hypothesis is proposed that all organisms that exist in the earth at present are genealogically related to each other, something which is asserted through Darwin's theory called "The Tree Of Life".

- Accuracy in the language used by biological theories is implemented. This means adjustments and adaptations of the concepts and their classifications, definitions, concerns, etc. Darwin replaces the essentialist traditional concept of species, defining them as "well marked varieties" (Darwin 1872, pp. 97-98). Such an idea is attributable to a group of individuals which are very much alike amongst each other and which do not differ in the term "variety". The historical relationship is so decisive that out of it, the species would be nothing else but "variety".

    Besides, the main theoretical and argumentative work of *Origin of Species* consisted of trying to demonstrate how apparently irrelevant observations referred to variation, competence and heritage, can answer questions which up to that moment seemed to be beyond the scientific field for its consideration or they had only been approached partially and restrictively. The argument generated by these questions centered more on the necessity to specify meanings than on proving the truth of these assertions.

- Another important novelty has to do with the explanatory pattern proposed by Darwin's theory. Such a pattern required the consideration of historical processes and the inclusion of contingent factors which do not only take part as "initial conditions" as in the prestigious nomological-deductive pattern but as fundamental components of the explanatory process. This type of explanation which incorporates the well-known aspects has been denominated "Darwinian historical narration" and it is characterized because it not only uses reasoning, laws or experiments, but also metaphors and analogies, that is to say, strategies which were not formerly scientifically accepted at all.

  Another difference of this explanatory scheme consists of considering that while the completion of explanations by laws is intrinsic, that is to say, more initial conditions or more laws can be added to the explanans, the historical explanations may be completed, also extrinsically referring to the aspects in context, without which it is not possible to give an account of the evolutionary phenomena which happen through time. This type of explanation has been argued from a hegemonic pattern of science, which considers the nomological-deductive explanation as the fittest to the scientific requirements.

- The above mentioned aspects also have an incidence on the importance of new methodological criteria; in fact, the Darwinian theory embodies original guidelines so as to classify relevant information on Biology since it must take into account the genealogy[2] that connects individuals, instead of placing them in a 'natural system' which is understood according to the accepted teleological perspective at that moment.[3] In order to get the proper information, Darwin is going to head precisely to empirical data which

---

[2] "We shall immediately understand why these characters (meaning the embryological ones) have such a great value in the classification: because the natural system is genealogical in its disposition. [...] Our classifications are several times evidently influenced by bonds of affinities." (Darwin 1872, p. 419)

[3] "So the natural system is genealogical in its order, as a genealogical tree, but the amount of modification that the different groups have experimented has to be expressed classifying them in what are called genera, sub families, families, sections, orders and classes" (Darwin 1872, p. 421).

nature offers putting aside the support that any theological explanation could give to the theory,[4] in order to establish reliable sources apart from suitable observations and experimentations.[5]
2. The theory includes, in what we call 'conception of *maximum*', certain thesis which have a negative impact not only in the scientific community, but in the general set up conception of the world at that time such as:
2.1. The introduction of the concept of 'natural selection' to which the configuration of one of the fundamental[6] mechanisms in the evolutionary process is attributed; it implies a high impact for the consolidated thesis of the invariability of the species.

> I have called natural selection or survival of the fittest to this conservation of the varieties, the variations and individually favorable differences and the destruction of the ones which are harmful (Darwin 1872, p. 116).

> I am completely convinced, that species are not only unchangeable,, but that the ones which belong to what is called the same genus are direct descendants of some other species, generally extinguished in the same way as the well-known varieties of any other species are their descendants. Besides I am convinced that the natural Selection has been not only the most important but the only means of modification (Darwin 1872, p. 58).

2.2. Darwin states the character eminently adaptable of the transformations that can happen to living beings against the strongly theological conception of the changes:

> Whatever the cause could be of each one of the slight differences among the descendants and their ancestors – and there has to be a cause for each one of them – we have grounds to believe that the continual accumulation of beneficial differences is the one that has

---

[4] "[...] the simplicity of the idea that each species was produced at first in an only region captivates the mind. He, who turns it down, rejects the true cause of ordinary generation with later immigrations and invokes the intervention of a miracle" (Darwin 1872, p. 374).
[5] "I have made so many experiments and I have collected so many facts that show on one side that an occasional crossing with and individual or different varieties increases the vigor and fecundity of the descent" (Darwin 1872, p. 287).
[6] Darwin also mentions other evolutionary mechanisms among which, we can mention the inheritance of acquired characters, the use and disuse and correlative growth.

originated all the other more important structural modifications, in relation with the customs of each species. (Darwin 1872, p. 185)

2.3. The transformations are fundamentally explained by the struggle for survival, without a pre-established aim. Although this principle, was already present in the thoughts at Darwin's time, he applies it to the field of inherent processes in the living beings.

> Let it be clearly understood that I use this expression Struggle for the existence in a wide and metaphorical sense that includes the dependence of a being from another – and what is more important – it not only includes the individual's life but also the success of leaving offspring. (Darwin 1872, p. 102)

2.4. The above-mentioned processes are not the result of a specific event but they rather respond to a calibration of variations

> [...] natural selection only works making good use of small successive variations, it can never give a high and sudden jump, but it has to advance with short and safe but slow steps. (Darwin 1872, p. 209)

> If it could be shown that a complex organism which did not develop through slight successive and numerous modifications existed, my theory would fail (Darwin 1872, p. 199).

3. The above mentioned characterization of biological changes allows us to have an integrated global look that shows the process of life in the tree-shaped distribution:

> The affinities of all beings of the same kind has sometimes been represented by a great tree. I think that this comparison expresses a lot of truth. (Darwin 1872, p. 157)

> In the same way as buds give rise by growth to new buds, and these, if they are powerful, ramify and over-push in all directions to lots of weaker branches, it is in my opinion, what has happened with the great tree of life, that with its dead and broken branches fills the earth crust and covers the surface with its beautiful ramifications in constant forking. (Darwin 1872, p. 158)

These theses were precisely the ones, which provoked important enough objections so that they determined their rejection by several biologists and especially from the theological framework, which impregnated the research work on the origin of life.

On the other side, even though the Origin of the Species contains an enormous wealth of data, the empirical support turns out to be insufficient to sustain, from an inductive point of view, the extent of the above – mentioned thesis due to the limitations of fossil records, and that generates a disbelief in the force of the theory. The greatest difficulties referred to the origin of variability and the mechanisms of inheritance, in relation to which it is fair to acknowledge that they had also been a matter of concern for Darwin himself.

The possibility of its identification with the materialistic ideology reinforces the negative attitude towards the theory, and this is a situation which becomes evident through different biologists' manifestations who do not recognize, for instance, the validity of the hypothesis of natural selection. This situation encourages the strengthening of neo-Lamarckism especially in America.

The pointed out distinction makes it more understandable the fact that the publication of the *Origin of the Species* marks a practically foundational example of Biology stating what P. Kitcher calls a new consented scientific practice, which means to acknowledge that the research in biology will remain marked by the above mentioned aspects in 1; it can be said that such aspects would mark the success of the Darwinian evolutionary theory.

But on the other hand, once the sparkles of the initial success are silenced, the aspects pointed out in favour the critical situation in which the theory fails at the end of the 19th century, acknowledged by the scientific community according to what publications of the time make evident and they refer to an "eclipse of the Darwinian theory".

The striking point is that several decades later, a revaluation of such a theory is produced, giving rise to an acknowledgement of the validity of their thesis for the new biological discoveries, something which consolidates and enlarges the validity of the theory. It is important therefore to identify the epistemological factors that influence in such a request.

It has to be taken into account that the second half of the 19th century, the evolutionary process is sufficiently accepted as a phenomenon which belongs to the development of living beings. Nevertheless, the evolutionary thesis does not exclusively belong to the Darwinism, since it is a part of different proposals, which consider it too. In the moment we are concerned about, other alternative theories were present in scientifically context, which from perspectives opposite to the Darwinian one, tried to offer a satisfactory explanation to such a process. To summarize, we point out the central thesis referred to the evolutionary process of the theories which sharing the scientific stage are valid in the last decades of the 19th Century.

1. Theistic Evolution: the variations are not hazardous but they are directed to the aims proposed by the design of the creator. Evidently, this proposal remains out of the possibilities of the scientific research.
2. Lamarckism: the evolution results from the process of inheritance of characters, achieved during the life of an organism as an answer to the environment through an impulse or an inherent vital force according to the usage-heritage relationship; it is possible the accumulation of bodily modifications created by a new pattern of conduct and adopted by an organism.
3. Orthogenesis: it conceives the evolution as a process consistently directed throughout a singular development by forces which operate in the same organism
4. Mutation Theory: it states that the evolution proceeds by the sudden apparition of new significative forms.
5. Natural Selection: it states that those individuals, which preferably survive and reproduce are the ones born with variations which confer them some adaptable benefit or advantage because of the demand of their environment, such variations would be the result of disturbances of the reproductive system which happened unfortunately.

Each of these proposals must individually consider questions referred to the characterization of the evolution explaining whether it is:
- A methodical process in which the groups move on through a regular pattern of development or an irregular process which generates diversities and tree-shaped ramifications.
- A process controlled by a (external) demand of the habit or by internal forces to the organism itself.
- A continuous process of accumulation of small regular changes or it rather occurs by a discontinuous appearance of completely new forms.

The theory of the Theistic evolution and Orthogenesis and the Lamarckism as well keep a line according to the secular tradition which states that the process of organic development should be well-arranged and controlled by inherent laws in life itself, though more of them have enough empirical data to support itself. Consequently, there is a tendency in the scientific community to face the challenge, which means to try new approaches for certain fundamental problems.

The situation being such that a new generation of biologists becomes conscious that the techniques of morphology and the field studies are such that they seemed to have reached the limit of utility and they are convinced of the necessity of an experimental tackle of these problems, which up to the moment, none of the above-mentioned alternative theories could offer. Weismann's works who re-discovered Mendel's laws in 1900, line up in this tendency, and besides he contributes with his studies on the chromosomes as a basic material of inheritance, something which would refute Lamarckism thesis on the transmission of achieved features. De Vries on the other hand distinguishes two types of variations, the ordinary one and the one which gives rise to great transformations in the organisms, introducing the concept of 'genetics mutation', which refers to a sudden change and without transition. William Bateson and other geneticists agree with this position.

The new approach turns out to be strongly hostile to Lamarckism, which had achieved an important space. In addition to this, genetics and the theory of mutation firstly appear as alternatives which would contribute to eclipse of Darwinism; they are put forward, at least potentially, as compatible with the thesis of natural selection. The strongly hereditary condition of the perspective, does not resist the theory of recapitulation but, on the contrary, is interested in the peculiar somatic characters with hypothetical genetic unities (by combination or creation of new unities); therefore, the thesis which considers ontogeny as the guideline of evolution is not accepted.

The Mutation theory is discussed from different points of view, among them we may mention the Biometric position headed by Karl Pearson who trust in the Darwinian thesis of natural selection focusing its attention in the quantitative or 'metrical' variations.

The contribution of mathematical instruments for elaboration of data leads to the appearance of the genetics of populations, making it possible the construction of theoretical patterns in which the action of selection over small genetic effects performed an essential role.

T.H. Morgan is also opposed to De Vries mutant theory. He is the one who has not only discovered that the genes are found and interbred in the chromosomes since 1910, but he has presented the evolution as a relatively gradual process in which the news genes who gave slight adaptable advantages spread in the population, though his concept of selection is a little bit simple and he rejects the thesis of the struggle for survival or elimination of the least fit as a mechanism or incorporation of new characters to the population.

Towards 1920, some genetics scientists started to admit that the formerly rejected thesis of the natural selection could play an important role in the explanation of evolutionary changes. Other Darwinian thesis strongly argued, such as the adaptable character of evolutionary changes and gradualism, start integrating little by little to the new proposals.

Among the field naturalist scientists, to whom the thesis of the section was still unacceptable from the reasoning of survival of non-adaptable characters, an important change was also produced a new way of selection obtained the help of several naturalists and allowed them to accept genetics as a basic component of the evolutionary theory.

In this way, between the decades of 1920 and 1930 Darwinian's theory has already regained a relevant position for Biology, and this is for instance, mathematically shown that Darwinian natural selection really produces important changes. Distinguished researchers such as R. Fisher, J. Haldane, S. Wright, etc., are able to shape a theoretical framework where Darwinian thesis are integrated with the contributions in the genetics field. Subsequently, Darwinism will not only constitute another biological theory, but it will shape the structural framework in which different disciplines of the biological field which seemed scattered up to the moment, are integrated without the founding elements of possible cohesion.

We can then admit that the revival of Darwinism is the result of two different processes of reconciliation:

1. Necessity to overcome the gap between Biometry and Mendelism: the discontinued variation was not compatible with natural selection (a new generation of biologists qualified in statistics to appreciate such a possibility was necessary).
2. Necessity to reconcile the heritage of Mendelian and Darwinian proposals with the preference for the Lamarckism or Orthogenetic mechanism of field biologists that rejected the germinal isolation

Naturalists were conscious of the absence of experimental evidence that backed up those theories. Because of the appearance of a new form of selection based on the genetics that agreed with the laboratory studies and it was also able to conduct a fieldwork satisfactorily, the support to the other theories quickly vanished.

The theory of genetics selection eventually succeeds because the reconciliation with Mendelism is inevitable if one wants to overcome the separation between biologists of laboratory and biologists of field and fundamentally because it is able to propose a model of science which accepts a multidimensional change (questions, explanatory schemes, instruments, etc) opposing to any other conception of science that may reduce one-dimensionally to progress. The complexity of the world, especially the biological one, and the methodological resources required a new scientific practice that was started by the Darwinism (Kitcher 1993, p. 44) and consolidated by Neo-Darwinism.

3. Conclusion

We believe that Neo-Darwinism contributes important conceptual novelties such as those which are related to the concept of species defined as a cluster of individuals which cross-breed among each other or they can do so giving fertile offspring, which opposes the Darwinian concept that turned out to be as an almost gnoseological construction. Another precise definition is that of 'fitness' or Darwinian adaptation which is understood as the capacity of a genotype to be represented in the following generations in comparison with other genotypes of the same population. This is a part of an important conceptualization which implies considering as unit of evolution not only the individual, but also those populations in which evolution is carried out.

It is clear that not only in the process of acceptance of Darwinism in the moment of the publication of the *Origin of the Species*, but in the revival that followed its decline, epistemological innovations contained in the proposal have had an important impact. Indeed, the advent of the theory involved the following aspects .

1. In the initial proposal
   - To scientifically explain certain biological processes without resorting to significant foundations.
   - To complement or substitute, by requirements of mechanicist tradition, the theological explanation of the biological phenomena.
   - To reduce the types of causes invoked as explanatory principles to the material cause and the efficient cause but expressed through relationships of eminently rational character and of empiral character
   - To incorporate the reduction of hazardous variations as unavoidable factors in the evolutionary process. (this question generates an important tension in the theory since it starts a conflict with the explanatory hegemonic pattern and with the deterministic conception of the world accepted in that moment)
   - To admit that evolutionary phenomena are explained through genealogical processes that refer to unique and unrepeated facts in the historical sense that can not be covered by legally-formed relationships
   - To settle a new 'scientific practice' (Kitcher 1993, p. 31) which will contribute with significant guidelines for the future biological research.
   - To shape a research programme that allowed us to integrate the historical process for evolutionary thesis and to be open towards the incorporation of the new contributions in Biology
2. In the Neo-Darwinism it is contributed in an efficient way to strengthen the methodological strategies which allow biologists to:
   - To identify the instances of Darwinian theory in the field of living beings assuming the evolutionary perspective which defines new observation fields

- To find ways of testing hypotheses which arise in the advancement of such instances, the genetic of populations can mathematically explain, how natural selection acts over clusters of individuals that grow geometrically, reducing overcrowding with the elimination of individuals which have not been benefited by favourable variations
- To explain special phenomena through Darwinian theory that from other theoretical frameworks do not turn out to be relevant, for example, the change of colour of moths in Manchester (Biston betularia)
- To develop theoretical consideration of the supposed processes in the Darwinian histories for example the ones related to hereditary transmission and to the origin and preservation of favourable variations.

The universality of the genetic code and the belief in the arbitrary combination of the code favour the hypothesis that all the organisms are relatives (Sober 1996, p. 82):
- To implement formal structures which make not only the quantification of evolutionary processes possible but the elaboration of patterns which allow us to predict progressive programmes.
- To configurate an epistemological paradigm which has gone beyond the field of biology, offering new elements which serve as a guideline to scientifically investigation.

Although the revival of the Darwinian theory within the inside field of Biology is a result of the processes of reconciliation between Biometry and Mendelism on one side and heritage of the Darwin and Mendel's proposals on the other with preference to the Lamarckism or Orthogenetics biologists of fieldwork on the other side, we understand that its importance has to do with its capacity of balancing different fields of biology giving unification to the discipline, besides it has been efficient to propose a model of science which accepts multidimensional changes (questions, explanatory schemes, instruments, and so on) opposing itself to any idea which may one-side dimensionally reduce its development.

The complexity of the world, especially the biological one, and the methodological resources required a new scientific practice that Darwinism started and neo Darwinian consolidated.

References

Ayala, F. (1982), "Darwin y la idea de progreso", *Árbor* 113 (442): 59-75.

Ayala, F. (1974), "The Concept of Biological Progress", in Ayala, F. & T. Dobzhansky (eds.), *Studies in the Philosophy of Biology*, London and Basingstoke: Macmillan, pp. 339-354.

Álvarez, J.R. (2000), "Analogías darwinianas: metáforas y/o conceptos", in Mora, M.S. *et al.* (eds.), *Actas del III Congreso de la Sociedad de Lógica, Metodología y Filosofía de la Ciencia*, San Sebastián: Universidad del País Vasco, pp. 331-341.

Bowler, P. (1983), *The Eclipse of Darwinism: anti-Darwinian evolutionary theories in the decades around 1900*, Baltimore: Johns Hopkins University Press.

Castrodeza, C. (1988a), *Ortodoxia darwiniana y progreso evolutivo*, Madrid: Alianza.

Castrodeza, C. (1988b), *Teoría histórica de la selección natural*, Madrid: Alhambra.

Castrodeza, C. (1999), *Razón biológica*, Madrid: Minerva.

Coleman, W. (1971), *Biology in the Nineteenth Century. Problems of Form, Function, and Transformation*, New York: John Wiley & Sons.

Darwin, C. (1872), *On the Origin of Species by Means Natural of Selection or the Preservation of Favored Races in the Struggle goes Life*, London: John Murray.

Dobzhansky, T. (1954), *The Biological Basis of Human Freedom*, New York: Columbia University Press.

Gilson, E. (1976), *De Aristóteles a Darwin*, Pamplona: Ediciones de la Universidad de Navarra.

Goodwin, B.C., Holder, N. & C.C. Wylie (eds.) (1983), *Development and Evolution*, Cambridge: Cambridge University Press.

Huxley, J. (1942), *Evolution: The Modern Synthesis*, London: Allen and Unwin.

Jacob, F. (1970), *La logique du vivant. Une histoire de l'hérédité*, Paris: Gallimard.

Jacob, F. (1977), "Evolution and Tinkering", *Science* 196: 1161-1166.

Jasdtrow, R., (1993), *Darwin, Textos Fundamentales*, Barcelona: Planeta-Agostini.

Kauffman, S.A. (1993), *The Origins of Order. Self Organization and Selection in Evolution*, Nueva York: Oxford University Press.

Kitcher, P. (1993), *The Advancement of Science*, New York: Oxford University Press.

Lewontin, R. (1978), "Adaptación", *Investigación y Ciencia* 26, noviembre de 1978.

Lloyd, E. (1995), "Objectivity and the Double Standard for Feminist Epistemologies", *Synthese* 104 (3): 351-361.

Martínez, S. (1997), *De los efectos a las causas*, Barcelona: Paidós.

Martínez. S. & L. Olivé (eds.) (1997), *Epistemología Evolucionista*, México: Paidós.

Mayr, E. (1997), *This is Biology*, Cambridge, MA: The Belknap Press of Harvard University Press.

Maynard-Smith, J. (1986), *The Problems of Biology*, Oxford: Oxford University Press.

Morin, E. (1973), *Le paradigme perdu: la nature humaine*, Paris: Editions du Seuil.

Olivé, L. (2000) *El bien, el mal y la razón,* México: Paidós.

Ponce, M. (1982), "Adaptaciones biológicas y explicación teleológica", in *Tercer Simposio de Filosofía*, México: Universidad Nacional Autónoma de México.

Regner, A.C. (1995), *A natureza teleológica do princípio darwiniano de seleçao natural*, Doctoral Dissertation, Porto Alegre: Universidade Federal do Rio Grande do Sul.

Richards, R.J. (1992), *The Meaning of Evolution. The Morphological Construction and Ideological Reconstruction of Darwin's Theory*, Chicago: The University of Chicago Press.

Ruse, M. (1986), *Taking Darwin Seriously*, Oxford: Blackwell.

Ruse, M. (1973), *The Philosophy of Biology*, London: Hutchinson University Press.

Sober, E. (1993), *Philosophy of Biology*, Boulder and Oxford: Westview Press.

Williams, E.O. (1975), *Sociobiology. The New Synthesis*, Cambridge, MA: The Belknap Press of Harvard University Press.

Wright, L. (1976), *Teleological Explanation*, California: University of California Press.

Chagas' Disease: History, Facts and Interpretations

César Lorenzano

National University of Tres de Febrero (UNTREF)

1. Introduction

At the end of 1999, Francoise Delaporte publishes a book in which he reinterprets, in a polemical fashion, the history that leads to the discovery of Chagas' disease.

The figure of Carlos Chagas, that of Salvador Mazza, as well as the Argentine-Brazilian community which studies this disease, are put into a perspective that clashes with the usually admitted versions. Delaporte does it by sticking outstandingly to the original texts, which he submits to conceptual and epistemic analysis.

For those of us who do not share his points of view, it is a challenge to take apart his well-constructed interpretive structure, which also compels us to revise all the pertinent bibliography. In the light of this discussion, very old materials acquire a fresh meaning. This is so because the differences we perceive between our position and that of Delaporte's go beyond precise indications or dissimilar interpretations of the writings.

They are associated with:
i. the perception of which aspects are relevant to Chagas' disease;
ii. the way in which scientific assertions are validated;

iii. and mainly, perhaps, with the epistemological conception, which deeply influences the structure of the historical account, the interpretation of the facts and the role that we assign *in this precise case* to the non-contemporary knowledge of the events that are being analyzed.

For this reason, and although we read the same texts, we see different things in them, and the stories we construct also diverge.

The present article is centered on the second part of his book, where he studies the "refoundation" of American trypanosomiasis, and the role of Salvador Mazza – whom he regards as a "fraud" – in that story.

Delaporte's central claims will be presented first and then, the points on which we dissent; the pertinent bibliographic material and the way in which it supports or questions his interpretations will be analyzed and interpreted later.

Finally, we shall see Salvador Mazza in his full stature, and we shall have a more exact picture of the historical, social, conceptual and epistemic mechanisms with which scientific knowledge is constructed, and which complement and rectify the mechanisms presented by Delaporte.

2. Delaporte's central claims

Delaporte argues against the generally accepted claim that, after a period of oblivion that lasted a dozen years, Chagas' disease is "rehabilitated" and its studies are resumed thanks to Salvador Mazza's and his collaborators' efforts.

This is not the case for Delaporte. For him, neither the studies are "resumed" nor is the disease "rehabilitated" after the attacks Chagas receives when people fail to acknowledge that this is an epidemic of maximum dimensions. According to Delaporte, the disease reaches its present dimension thanks to the discovery made by Cecilio Romaña (1935) – who was a young Argentine doctor and scientist at the time, and a disciple of Mazza's – of the sign that bears his name, and which consists of a conjunctivitis with unilateral palpebral edema, accompanied by regional adenopathies. This sign, which helps to diagnose the acute stage of the disease, enables to increase outstandingly and in a short time, the amount of diagnosed patients and consequently establish its real epidemiological importance. Mazza, who attributes the sign to Chagas and to himself as his follower – says Delaporte – is a *fraud*.

But the question here is not only that Romaña rehabilitates Chagas' disease. He "refounds" it, because he epistemologically places it on another field. In the first place, he categorically separates it from the endocrine disturbances, among which Carlos Chagas had situated it when he stated that the trypanosome affects mainly the thyroid gland. From Romaña's works on, we see American trypanosomiasis as a parasitic and not as an endocrine disease. Secondly, in Romaña's sign, the portal of entry of the infection – the conjunctiva – coincides with the clinical symptomatology observed, since the conjunctivitis is caused by the contact with the feces of the triatome, which serve as a vehicle for the infecting forms of the trypanosome – in order to enter the organism through there.

We have hitherto referred to the characterization Delaporte makes of the "refoundation" of Chagas' disease. Let us see now its different aspects, and the arguments which prove or refute these claims.

3. The structure of the disease

American trypanosomiasis or Chagas' disease is a disease caused by a parasite, the *Trypanosoma cruzi*. When studying it, we have to consider at least three aspects:

The first one is the *parasitological*. In it, we study the natural evolution of the parasite which, in this particular case, has a double cycle: on an insect of the Triatome genus – the *vinchuca* in Argentina, the *barbeiro* in Brazil –, and on intermediate hosts, mammals, among them, man.

The second aspect is the *clinical* aspect, in which it is considered an illness. As in many infectious and parasitic diseases, there is a portal of entry to the microorganism, a primoinfection, an acute stage of the illness and a chronic period, which is reached after a period of latency.

In each one of these stages, the clinical signs are closely linked to anatomophysiopathological alterations which are, in turn, accompanied by immunological manifestations.

But the disease is not only an individual event. It is also important as a social process, i.e. its incidence in human populations and the environmental conditions in which it develops. This is the third aspect of the disease, the *epidemiological*. If the first two are important to diagnose the illness, this third aspect is crucial, because the prevention of the disease depends on its correct interpretation.

The simple enumeration of the complex structure of the disease gives us a clue to the disagreements with Delaporte's historical version. We notice that in it, the emphasis is laid on the acute stage of the clinical aspects of the illness, at the expense of the chronic stage and certainly, of the epidemiological aspects.

As we shall see later, if these three aspects are taken into account, if they are assigned the importance they really have in the perception of the disease, the historiographic perspective that Delaporte constructs changes in order to locate its historical construction on another ground.

In it we can see a certain continuity in the process, and not so much a complete epistemological break; we can see refinements and corrections of an already established conceptual frame, and not a totally new construction. Thus, Carlos Chagas' role continues to be central in this story, and not the mere ground for refutation and the later refoundation of a disease.

4. The contributions of Carlos Chagas

In order to evaluate Carlos Chagas' work in its right dimension, we are going to mention briefly the main contributions he makes as well as the obstacles he encounters.

He finds an infecting agent, the *Tripanosoma cruzi*, in an insect, the *barbeiro*. He describes its life cycle on the insect and on the human host. He describes a portal of entry and an infecting mechanism. He finds acute phases of the illness, certifies them through positive identification of the trypanosome in the patients' blood, and manages to reproduce on animals which have been inoculated with that blood, the same lesions found in patients. In his anatomopathological studies, he finds chronic lesions in various organs. He studies the geographical distribution of the *barbeiro,* of the trypanosome and of the patients. He determines in which environmental conditions the disease develops, and the sanitary measures taken to prevent it. As we know, the strong opposition this disease generates among the powerful people in Brazil is connected with the social criticism of the living conditions of the people who suffer it. This is what happens to Virchow when he realizes that these are the underlying cause for the Typhoid fever epidemic in Silesia, and also to Mazza, in the north of Argentina.

According to what we see, his studies cover all the aspects relevant to parasitic diseases. Nevertheless, not all his findings are validated by the scientific community, made up by doctors, parasitologists and epidemiologists.

This does not mean that he is mistaken about all and every one of the aspects of the disease. The *barbeiro* is the carrier of the trypanosome which infects man, undoubtedly provoking in him – and this is certified by the fulfillment of Koch's postulates – an acute illness in the (paradigmatic) cases through which he initiates the knowledge of the disease. He very rightly describes most of the many anatomopathological lesions of the illness which are today recognized as such – those corresponding both to the acute and to the chronic stages. He adequately locates the geographical distribution of the *barbeiro* and of the trypanosome, and their epidemiological importance.

However, mistakes are pointed out to him very soon. Many are derived from the difficulties which are typical of the study of infectious diseases, others from the special historical and geographical circumstances in which he does his work, which enable him to achieve his greatest discoveries but, at the same time, veil other aspects.

As regards the specific study of the parasite, its life cycle and its inoculation in human beings, he makes mistakes which are strongly pointed out to him. The trypanosome does not reproduce sexually, it does not have a phase in the human lung, and is not transmitted through the bite of the insect, as he declares.

It is not simple to follow all the evolutionary line, if what the researcher has at hand are only snapshots, frozen moments of the cycle of the parasite which he has to organize, imagining the transition between them until they form a continuum. Is it strange, then, that he should use evolution models already established for other parasites, in order to devise the cycle of the trypanosome? Even if he does not take them mechanically, even if he adapts them to what he finds, he cannot help having gaps in his interpretations, he cannot help erring. If malaria is transmitted through bites, and if he finds trypanosomes in the salivary glands of the insect, what else can he think but that the *barbeiro* transmits the disease through bites?

Chagas observes and interprets according to the knowledge of his time. That is why he is mistaken. And the same cause that leads him to discover the disease – the profusion of infectious and deficiency diseases of the area under investigation – also leads him to error. It enables him to direct his attention to the *barbeiro* and find the trypanosome while he is studying malaria. But at the same time, and in addition to his disease, he finds a lot of other ailments in his patients. As he finds it difficult to separate them, he interprets the pulmonary form of another microorganism as typical of the trypanosome.

Nevertheless, this is a minor mistake. What is more serious is the overlapping of the territorial domain of the trypanosome and of the disturbances of the thyroid, at a moment in which the origin of goiter and thyroid insufficiency could still be discussed. This leads him to privilege the alterations of the thyroid, above all the other anatomopathological alterations he finds, as characteristic of the disease. When he places stress on them, he turns the parasitosis into a mainly endocrine disease (parasitic thyroiditis).

The errors about the trypanosome cycle and about the mechanisms of infection are corrected by Brumpt and by a group of young Brazilian scientists who work with Chagas at the Oswaldo Cruz Institute. Neither Brumpt nor the Brazilian scientists doubt Carlos Chagas' enormous contribution to the knowledge of the *Trypanosoma cruzi*, nor do they dispute his paternity of the disease.

As regards the preeminence of the lesions of the thyroid, Kraus *et al.* confirm very early (1915, 1916) the existence of an insect similar to the *barbeiro*, the *vinchuca*, scattered over a great geographical extension in the north of Argentina, which is parasited by the *Trypanosoma cruzi* although there is no endemic goiter in the area. Neither do they find *Trypanosoma cruzi* in the blood of evident goitered people and cretins, nor do they obtain them from experiments on laboratory animals. The situation has the appearance of a classic refutatory experience, and it is thus viewed by Chagas' opponents, who conclude that the chronic illness does not exist, although perhaps the acute illness does in just a few confirmed cases.

Delaporte comments – and this is one of the many points on which we disagree – that what makes historians say that Chagas' resistance against his opponents (when he states that he is right, and that the disease exists) is justified, is a retrospective illusion. It is also an illusion to say that Kraus was right against him. This is so because "The essential is somewhere else: at that time nobody is in a position to separate the morbid entities, either different or overlapping. The pure form of Chagas' disease cannot be perceived for the simple reason that it is not constituted" (p. 131).

5. Romaña and the refoundation of the disease

On reaching this point, Delaporte presents all the preconditions he needs to support his main claim: that Romaña refounds Chagas' disease when he discovers the sign that bears his name.

The Delaporte's move is the following: up to that moment, Chagas' disease is a parasitosis, but not a clearly defined disease. Only with Romaña's sign does the parasitosis become the "pure form of Chagas' disease", that which combines the presence of parasites with its unequivocal (pathognomonic) sign, that which brings together the portal of entry of the parasite into the human body with the inoculation mechanism: the insect's feces carrying infecting forms of the trypanosome, entering the body through the conjunctiva.

Seen in this light, up to Romaña, Chagas' disease comes under the domain of parasitology (as the discipline that studies parasites in general) and not under that of infectious and parasitic diseases, or perhaps under that of harmless parasitoses, as Chagas' opponents first and Mazza's opponents later declare.

Of course, Delaporte takes good care not to say that the general signs of illness suffice to define a clinical entity. It is enough to remember that many of the viruses do not present any other symptomatology, and that they are much less spectacular at times than that of Chagas', in order to be considered as clearly defined illnesses and not as the carrying of a virus by a healthy individual. If he said that, he himself would start demolishing his claim. Neither does he mention all the lesions Chagas describes – apart from the thyroid lesions – and which cause cardiac, nervous, digestive and other symptoms.

In spite of Delaporte, Chagas diagnoses the illness *clinically* just by looking at the patient. Brumpt quotes him (1913, p. 187) and expresses: "The patient's face presents a distinctive swelling which enables us to suspect the illness at a distance."

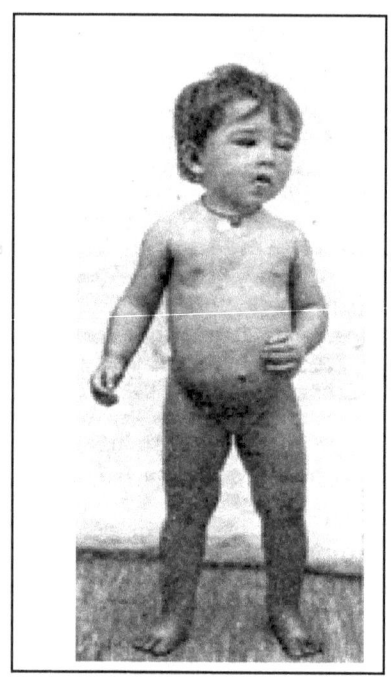

In the picture Brumpt uses to illustrate Chagas' disease in its acute stage (Fig. 113), taken by Carlos Chagas and given to Brumpt by Couto, we readers are also able to diagnose the illness at a distance. We are right: "the child has a great number of parasites in the peripheric blood", Brumpt tells us in his book.

But then Delaporte says what Chagas sees is a hypothyroidism (although he may have parasites in his blood; it is precisely a parasitic hypothyroidism) and he cannot but see it in this way, because so far, the "pure form of the disease" – unilateral palpebral edema, adenopathies, without hypothyroidism – does not exist (conceptually). "Pure form" means with no other overlapping disease.

The circle has closed and the argument is merely definitional: Chagas cannot diagnose Chagas' disease because Chagas' disease has not been constituted yet. Or worse even, if the argument is epistemic – and perhaps this way of understanding it is the one that does more justice to Delaporte's words – we are dealing with two different diseases, the first of which (Chagas') does not exist, since it is (erroneously) endocrine, and the second one, which (deceitfully) has the same name, starts with Romaña. (Actually, if we go to extremes, it is Romaña's disease.)

By holding that this "pure form" is the characteristic of the disease, Delaporte loses sight of the chronic forms (if we follow his reasoning to the letter, they are not pure forms: they affect a particular organ, and their signs overlap with those caused by the parasite.). If he did not do that, he would discredit the very core of his claim, since he would see that Romaña's sign identifies only the acute forms and is, therefore, only *partially* responsible for the revival of Chagas' disease.

6. The chronic illness

We had mentioned that one of the essential aspects needed to know about an infectious disease is that which refers to its chronic stage (if it has one).

In the case of Chagas' disease, it is central. It is through its chronic stage that one can know about it. If the disease were restricted to its acute stage, and this consisted only of Romaña's sign, there would be two possibilities: that after this stage there would either be a complete recovery or that the parasite would harmelessly remain inside the body. Exactly what Chagas' critics tell him. The acute stage is, at least in Argentina – where Romaña does his research – of benign course. So much so, that frequently it either goes unnoticed or it is denied importance.

It is the chronic stage of the disease, its pathological and epidemiological weight, which makes it interesting and justifies the enormous preventive campaigns aimed at eradicating it.

From the clinical point of view, as it is known since Kraus' time, chronic Chagasic patients do not suffer from any thyroid alterations. They have stopped mentioning them in all texts and they have been replaced (mainly) by cardiac and also by digestive conditions. Chagasic myocardiopathy is the great clinical manifestation of the disease and, to a lesser degree, megacolon and megaesophagus (Bennett & Plum 1996, p. 1899: "The chronic manifestations of the illness develop years after the initial infection, as a myocardiopathy with defects in the conduction or with esophagus or colon dysfunction – megasyndromes –").

As far as epidemiology is concerned, the number of patients who suffer heart or digestive disorders is not, by any means, small. The *Trypanosoma cruzi* is responsible for the greatest pandemic of parasitic origin in Latin America and third pandemic in the world, after malaria and schistosomiasis (WHO 1995, p. 125):

> About one hundred million people – one quarter of the total population of Latin America – are at risk of contracting the disease. It is estimated that between sixteen and eighteen million people are infected. The illness usually starts as an acute infection during childhood. It can last, at the most, two months, and it is followed by a slow, chronic inflammatory process which damages the autonomous nerves and the heart tissue in about a quarter of the people infected, a condition which may cause heart failure and premature death at middle age. In about 6% of the people infected, the autonomous nervous system of the intestine is affected, leading to disturbances in the peristalsis and dilation ("megacolon" and "megaesophagus"). In 3% of the cases the peripheral nervous system is involved.

In Brazil, Dias (1979) reports that in Minas Gerais, where Carlos Chagas conducted his research, Chagasic cardiopathy affects about 40% of all the infected adults.

Undoubtedly, American trypanosomiasis really has the importance Chagas assigns to it, and during its chronic stage it presents heart disturbances which were studied by him.

It is not by chance then, that Romaña (1963, p.64) says:

> Although Chagas' disease is, during its acute stage, a serious infection due to the deaths it causes among children, its true sanitary importance lies in the chronic forms and particularly, in the cardiopathy manifestations it develops

and he adds:

> Many of the parasitized individuals present morbid pictures of chronic evolution, among which only Chagasic cardiopathy could defini-

tively be individualized and described. This is the only undisputed syndrome of the vast clinical panorama originally pointed out by Carlos Chagas; apart from cardiopathies, only some nervous and gastrointestinal syndromes are attributed today to the *Trypanosomiasis cruzi*.

Is it possible to hold, as Delaporte does, that Chagas' disease is "refounded" because it is simpler to identify acute patients after Romaña? Or should we rather locate the refoundation when the chronic phase is identified, and we can visualize its clinical and epidemiological relevance?

The simple posing of this question centers the query about the revival of interest in Chagas' disease, in the resignification of Chagasic cardiopathy which takes place around the 30s.

7. Chagas' disease, a heart disease

The central historiographic question that we are now posing is when exactly Chagas' disease becomes mainly a heart disease and starts having the epidemiological weight we have just mentioned.

In his first articles, Carlos Chagas already mentions heart lesions he has identified in Chagasic chronic patients through careful anatomopathological studies which show trypanosomes in the myocardium. In his mature years, Romaña (1963, p. 64) comments on the important role played by Chagas in the study of the chronic manifestations of the illness:

> The symptomatology of chronic cardiopathy was outstandingly studied – for those days – by Carlos Chagas and his collaborators Eurico Villela and Evandro Chagas. The clinical research they did provided the basic elements for the diagnosis, prognosis and treatment, and subsequent studies – thanks to the modern electrocardiographic techniques – only helped to specify the reactions and symptoms. Thanks to the epidemiological studies, they were also able to show the importance and extension of the ailment.

Crowel (1923), who studies Chagas' disease from the anatomopathological point of view at the Oswaldo Cruz Institute for about four years, expresses: "In the chronic cases, the parasite affects mainly certain systems, thus laying the foundations for Chagas' classification of the chronic cases into cardiac, nervous and pluriglandular forms." (p. 426)

A few years later, when Magarinos Torres (1935) studies the myocardium in fifteen chronic cases of Chagas' disease, he finds it affected by chronic myocarditis. He states:

> there is myocarditis in continual and progressive evolution in Chagas' cardiopathy, because the Trypanosoma cruzi does not confer such a degree of *immunity in the individual who contracted a first infection as to prevent a new infection followed by the multiplication of the parasite in the tissues and a small number of these in the blood; the most the individual acquires is an allergy state*. (p. 914; we are responsible for the italics)

Apart from Chagas himself, and Magarinos Torres, researchers like Vianna and Crowel also found trypanosomes in the myocardium of chronic patients.

In the Fifth Meeting of the Argentine Society of Northern Regional Pathology (Sociedad Argentina de Patología Regional del Norte) which takes place in 1929, Salvador Mazza presents a chronic heart form of Chagas' disease confirmed by inoculation on a lab animal (small dog). This article is later published in 1935.

Romaña (1934 b) finds the first two cases of chronic Chagasic cardiopathy, published by the Study Mission of Argentine Regional Pathology (Misión de Estudio de la Patología Regional Argentina– *MEPRA*), the institution that Mazza founds in Jujuy as a branch of the Institute of Surgical Pathology (Instituto de Patología Quirúrgica) of the School of Medicine of Buenos Aires University. A fact that Delaporte barely mentions and to which he apparently attaches no importance. However, it is a step forward – and not a minor one – in the historical construction of the chronic stage of Chagas' disease.

Mazza himself presents together with Jörg (1935, pp. 229-230) an interesting paper on the anatomoclinical periods of trypanosomiasis, in which they characterize the acute illness as a primoinfection, consisting in a primary complex with satellite adenopathy, local swelling in the site of the inoculation and the infection spreading to the different viscera (heart, liver, meninges, etc). With respect to the chronic illness, it is defined by lesions such as sclerosal chronic myocarditis, myositis, splenitis. We must highlight the fact that they do not mention the existence of thyroid disturbances either in the acute or in the chronic illnesses. In this article, the separation between Chagas' disease and hypothyroidism or goiter is complete, and the organ which is mainly affected is the heart.

Chagasic myocardiopathy slowly starts to be regarded by researchers as one of the central forms in which the illness manifests itself. Here is where its revival (or "refoundation", if we follow Delaporte's terminology) starts. In it, the task of the *MEPRA* – and certainly of its founder, the tireless Salvador Mazza – is central. He demonstrates the existence of parasitized *vinchucas* in practically all the Argentine territory, with an epicenter in the north of the country. Until his death in 1946, he describes nearly one thousand four hundred diagnosed cases, the biggest casuistry ever gathered, and which shows the epidemiological importance of the disease. Thanks to Mazza, this is not anymore that patientless parasitosis Chagas is reproached for.

Already in the mid 30s (let us remember that Romaña presents his sign in 1935), the specific alterations in the electrocardiogram are a major sign of the illness, both in human patients and in lab animals, to the extent that its verification suffices to hold that the parasite affects the myocardium.

In 1938, doctors J. A. Aguirre and Clodomiro Jiménez present a paper in the Sixth National Congress of Medicine (Cordoba, Argentina) about 168 chest tele-X-rays taken to Chagasic patients, in which they prove that 86% have heart lesions which result in an increase of the heart silhouette, which is characteristic in its shape and evolution, because the heart cyclically changes its size.

A chest tele-X-ray or an electrocardiogram are enough for physicians to unmistakably diagnose Chagas' chronic disease.

Chagas' disease acquires, then, the features it has today. The reformulation of the disease has been completed in all its facets.

8. The mistaken question

Delaporte bases his text on classical questions of traditional historiography, and which are summarized in the exclamations: "What is discovered? Who did it? When?" Taking for granted that Chagas' disease is "refounded", he answers that Romaña does it, in a tight cluster of three articles, in which he establishes the sign that bears his name. A certain man, at a certain moment, a unique event.

When we see that what is most important in Chagas' disease lies in its chronic phase, and that Chagasic cardiopathy is based on the research done by more than one scientist who, in turn, base theirs on the original studies of Carlos Chagas, we start suspecting that the questions are not the right ones. They belong, as we think, to a theoretical sphere in which the social and collective aspects of science are neglected – or unknown – since it is only from an individualistic perspective that asking oneself who discovers what and when, acquires meaning. A romantic vision of science in which neither heroes nor villains are missing.

It is not only that he may be mistaken in establishing the moment of the resumption of the studies. The point is that there is neither an exact moment of historical inflexion nor a hero that takes it upon himself the heavy task of gestating something absolutely new.

When a historical and social conception of science is adopted, we perceive that it develops thanks to the – unequal, perhaps – contributions of a community of researchers who take as the purpose of their work, the opening contributions of those who explore a certain field of knowledge for the first time.

It is true that, until his death, Carlos Chagas held that what is essential in his disease are the thyroid lesions. But he was not being irrational when he thinks that, regardless of the supposed refutation. In the first place, his claim of parasitic thyroiditis is in accordance with the science of his time, which still believes in the microbial or genetic etiology of the thyroid disturbances. Secondly, he can claim (ad hoc) that the infection conditions change due to the warm climate in Argentina, preventing parasitic thyroiditis from appearing. In addition, there is important evidence that shows that Brazilian goitered patients are infected by the trypanosome; all or most of them present a complement fixation reaction positive for the illness, the Machado Guerreiro reaction (1913).

Do errors play down the importance of Chagas' work or are they part of a process in which this author took the first transcendental steps with which he set the agenda for the research of the most important parasitic disease in this part of the world?

It is not necessary to be an orthodox Lakatosian or Kuhnian to agree with Lakatos or Kuhn that every research encounters, from the very beginning, a number of questions to be answered. Nevertheless, it is from there, from the opening researches, that the gaps it leaves start to be filled.

It is that characteristic of unfinished knowledge that has made advancement possible, since research precisely consists of moving forward through the roads opened by the first, solitary, paradigmatic studies in each field of knowledge. Of course, this also demands thinking that the construction of knowledge is made by a scientific community and not by isolated researchers. Nowhere else like in Chagas' disease does this collective character of science become evident.

Without Chagas' work, the structure of knowledge which those who continue studying American trypanosomiasis base their studies on, would not have existed. Neither would its rebirth have existed without his stubbornness in maintaining its existence. If he had admitted he was totally wrong, as his opponents want, Mazza would not have believed him and would not have resumed his studies on Chagas' disease, and specially above all when the research done by Kraus – his friend and colleague – excludes the existence of the disease in Argentina.

9. Romaña and Mazza

The question about who discovers what and when, makes Delaporte belittle Mazza's (the fraud, he says) role in the resumption of the studies on Chagas' disease. In order to do this, he reexamines an old discussion between Romaña and Mazza in the light of the articles written by E. Dias, from which his central claim and his arguments seem to stem.

The problem, just as Delaporte presents it, is about the priority of a discovery, of that which we call Romaña's sign today.

While Romaña and Delaporte talk about discovery, Mazza insists that there is no such thing; the unilateral bipalpebral edema was already well-known to Chagas and obviously to himself, his loyal disciple.

Although we have already mentioned the epistemic difficulties in establishing this priority, we are going to follow all the history of the controversy carefully, expecting the revision to cast some light on it. I apologize to the reader for the abundance of quotations, but they are necessary to make a fair analysis of the facts.

We will see later that both Mazza and Romaña agree on the central points of the controversy, above their personal confrontation.

10. Carlos Chagas' article

We will deal first with the article in which Carlos Chagas gathers all the observations he made since 1909, when he discovers American trypanosomiasis. We will use Salvador Mazza's version. He translates the text and contributes with an introduction as well as with a few notes in which he refers to the interpretation of the cases and of their photographs. This consists of 29 observations, which make up the whole of his casuistry.

At the beginning of the article, Chagas describes (p. 12) the aspect of the acute cases: "the facies of an acute case of trypanosomiasis is almost always typical: flushed and swollen aspect; subcutaneous infiltration of all the face, with swollen eyelids, half-closed eyes, thickened lips and the tongue sometimes thick and furry", referring later to the rest of the signs, among them, adenopathies.

Here, we do not see him talking about the unilateral palpebral edema: this is a general description, and it refers to the – usually serious – cases he diagnoses.

If we summarize the description of the eyes in each one of the cases, we see that there is not a constant descriptive pattern which can make a perfect comparison possible. Thus, he mentions the existence of bilateral swollen eyelids, with or without face puffiness; keratitis with double conjunctivitis, facial swelling, without mentioning the eyelids, general infiltration, which he very often mentions as myxedematous.

Only in observation 16 does he mention the edematous swelling of only one eyelid (the right eyelid), and in observation 28 he makes reference to a pronounced conjunctivitis of the left eye and keratitis of one of the eyes, signs to which he adds, in case 6, a pronounced infiltration of the face.

Let us point out that in his casuistry, Chagas expressly indicates that in at least two cases, 16 and 28, there was unilateral ocular symptomatology, and in the latter case, with conjunctivitis.

If we now look at the photographs that accompany the observations, we will see that case 16 presents all the characteristics which, at first sight, we are used to associating with Romaña's sign. Unfortunately, there is no photograph accompanying case 28.

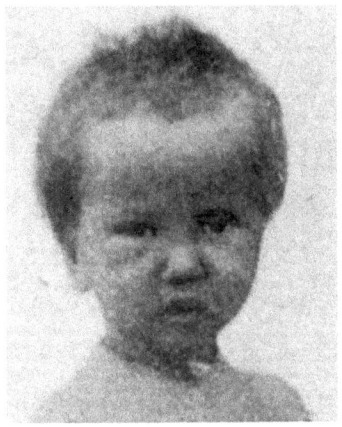

Case 16: Undoubtedly, the patient presents unilateral edema of both right eyelids.

Surprisingly enough, on examining the rest of the photographs – which belong to the cases in which Chagas does not mention unilateral palpebral edema – we verify that at least case 6 unquestionably presents Romaña's sign.

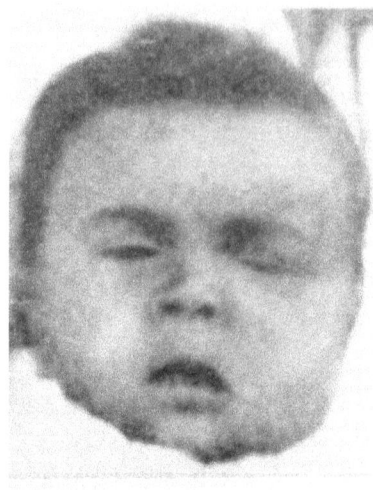

FIGURA 2

Case 6. We are faced once more with a Chagasic patient with the "sign of the eye", later called Romaña's sign.

It is probable – although there is some degree of doubt – that in the figures illustrating cases 5, 18, 22, 26 and 27, one of the eyes is more inflamed than the other.

In relation to case 16, which has all the appearance of Romaña's sign, Chagas (p. 33) comments: "Ten days ago, remittent fever appeared. At the same time, a small reddish papule was observed on one of the eyelids, with edematous swelling of the same and of the corresponding orbital rim", thus indicating that the portal of entry was a skin bite. Perhaps this is what Romaña is thinking about (1935) when he indicates that although this is rare, we should not discard that the infection entered through the skin and over all, through the more delicate skin of the eyelids.

There is no way for us to know if what we see in the figures today and read in Chagas' text is what his Brazilian disciples saw and read. Maybe not, because it would seem that E. Dias and E. Chagas can see the unilateral palpebral edema only after Romaña shows it to them, and it is for this reason that they ask it should bear his name (or they did so because it meant a confirmation – as we shall see later – of their own discoveries).

In any case, it is highly probable that Mazza sees them the way we do. Otherwise, we cannot explain how he does not show any surprise when he very closely supervises Romaña's research, supports it and disseminates it in his review.

11. Mazza, the fraud

A proof of what we have just said is a case of acute Chagas' disease found by Mazza in 1927. Although this is not published at the moment, it constitutes one of the examples that he uses in his dissemination talks for years. He gives the photograph away to Niño, who presents it – together with other cases – in his Ph. D. thesis in 1929.

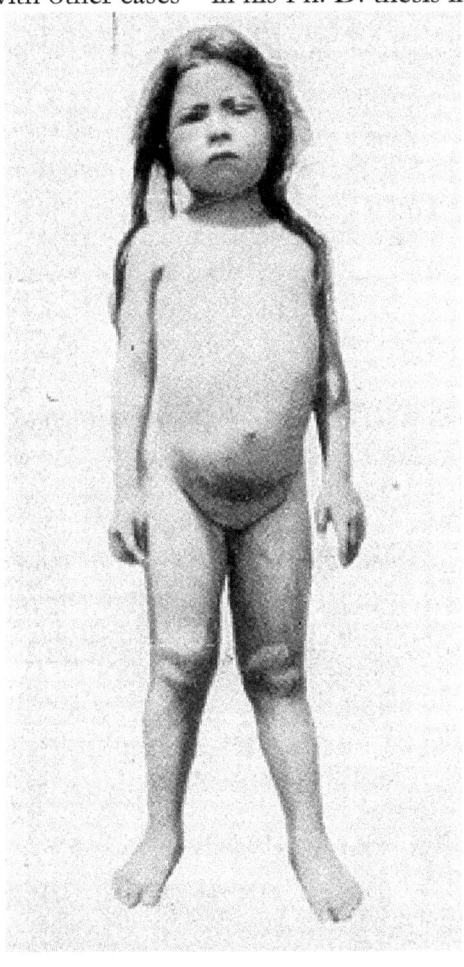

When we look at it, we find ourselves faced with a case which is similar to those of Carlos Chagas. We can see that the patient undoubtedly presents a unilateral palpebral edema, and that the diagnosis made based on the symptomatology is correct.

The point made by Delaporte is that, at that moment, Mazza does not mention the unilateral palpebral edema among the symptoms the patient has, and neither does Niño, who refers to an "edema of eyelids and extremities" (1929, p. 202). Only after Romaña presents it as pathognomonic and not before, is Mazza able to see – retrospectively – in his patient the sign as such. He did not discover it, but he was taught to see it. On saying that he saw it before, he – according to Delaporte – is committing fraud, since he aims at superseding Romaña as its discoverer.

Undoubtedly, we see Romaña's sign in Mazza's small patient. Nevertheless, it is true that in 1927, people do not yet know of the enormous frequency of occurrence of the unilateral palpebral edema and its consequent importance for diagnosing the illness, something which, as we well know, does not prevent him from diagnosing it correctly.

As we shall see later, Romaña himself does not mention, in the first seven cases, that the sign is pathognomonic and makes, likewise, a retrospective acknowledgment of the sign.

Is it fraud or a psychological mechanism that presents what is known today as if it had been known forever? We do not know what went on inside Mazza's mind. We can only assert that he was the first to find a case of Chagas' disease with unilateral palpebral edema in our country and he diagnosed it correctly, although only after 1935 does he start thinking, together with Romaña, that this is the main manifestation of the acute illness.

Beyond all ad-hominem interpretations of Mazza's personality – proverbially intolerant – and of his intentions, we will proceed with the historical reconstruction of the discovery and with the interpretation of the events.

Between 1934 and 1935, Mazza and Romaña publish a series of articles at the *MEPRA* alone, together or with other authors, which take up almost all the pages of the Review and which reveal two authors who support each other in establishing the validity of Chagas' disease in our country, the place where the discredit of the Brazilian researcher is initiated. The articles show how they trust each other, which is in keeping with their condition of mentor and disciple. They sign the last article together in 1936. From then on, they work separately.

Let us re-read them, expecting to find in them a clue that should enable us to understand and resolve their differences.

12. Romaña's and Mazza's articles

The article that initiates the series (Romaña 1934a) includes the verification of the first acute forms of the illness found in the areas of Chaco and northern Santa Fe. This is extremely important, since it takes place in the area where Kraus did not find any pathology at all.

Some previous works done in common establish a solid mutual relationship (Mazza & Romaña 1931a, b, c, 1933).

It is worthwhile paying some attention to this first article in order to evaluate the nature of their relationship:

> Since 1930 we have been doing research under the guidance of professor Salvador Mazza, aiming at establishing the real nosologic role of American trypanosomiasis or Chagas' disease in the north of Santa Fe and Chaco, up to now virgin areas as far as studies of this type are concerned, but which *have* been carried out for a long time in the north west of the country [...]. We, country doctors, could have advanced very little in our task, or it would have remained unfinished, had it not been for the generous help of professor Mazza's, our constant moral as well as technical supporter, true master of energy and industriousness and whose ability and advice were always at our service.

The situation is clear: even before graduating in 1931, Romaña is introduced by Mazza into the studies of Chagas' disease; he signs his first works with him, probably under his complete guidance. Mazza, who is permanently looking for contributors for his research, allies in the epidemiological fight against the transmissible diseases of the area, does a lot of teaching among the doctors and other social agents who might help him; he gives lectures, does research in the area, founds scientific societies in all the provinces. Young doctor Romaña is one of those doctors Mazza interests in the diagnosis and treatment of regional diseases; possibly one of his most talented disciples, with whom he collaborates and whom he respects, at least – as we shall see later – until 1936. In turn, it must have been extremely refreshing for Romaña to find an opportunity to channel his interests, lost as he was, in a village in the hinterland of the country.

Romaña comments that the existence of *vinchucas* with *Trypanosoma cruzi* and of wild animals, all that (Romaña 1934a, p. 5):

> led us to think that human cases, either acute or chronic, should be found in the area [...]. Once engaged in the search and diagnosis of human cases, it was not difficult for us to verify the infection in two children from the area with trypanosomes in the circulation and in two adults with cardiac lesions typical of the chronic form of the illness.

He describes the eye lesion of the first case, the eight-year-old Isabelino Martínez (p. 5-6) thus:

> In the early morning of the 1st. February 1932, the child wakes up feeling as if he had been bitten by an insect and with intense discomfort in the left eye [...]. That same afternoon the family noticed that the eye was swollen, and it was even more so by the following day [...]. The edema continued to be intense and almost painless on the following days, and two or three days later, his mother saw that it had also appeared in the right eye [...]. When I examined him, I noticed edema in both eyes, much more noticeable in the left eye, whose palpebral and ocular conjunctivas were red.

In the account, Romaña does not seem to consider that he is faced with something new (the swelling of only one eye) but with something common and familiar, which enables him to diagnose the disease easily.

The patient travels to Buenos Aires (on Mazza's advice), where he is examined by doctor Niño, Mazza's collaborator, and his case studied at the National Clinical Hospital (Hospital Nacional de Clínicas).

Romaña mentions (p. 13) that "doctors Acuña and Puglisi presented a fragmentary case history in the session of the 28th. June 1933 at the Argentine Pediatrics Society (Sociedad Argentina de Pediatría) and published it in the No. 5 issue of the review *Archivos de Pediatría*."

Romaña obviously did not imagine at that moment that he could be robbed of the priority of his discovery; in his view, Acuña and

Puglisi only contribute to disseminating the knowledge of Chagas' disease in the country and that was, in itself, valuable.

Professor Mazza studies the histology of animals which have been inoculated with the patient's blood (p. 14).

Finally, in the "Commentaries" with which he finishes the article, he expresses (p. 15): "The symptomatology and the interesting facts narrated by the mother, made me suspect from the very first moment that etiology, which was confirmed on finding trypanosomes in a big drop of blood."

Once again, the text reveals that the diagnosis is in accordance with the previous knowledge of the symptomatology of the illness, which is not new.

He adds: "That the infection has entered through the conjunctiva is a very well-based supposition, and it could have resulted from a bite or from the feces of *vinchucas*; it is interesting to find a preauricular ganglionar reaction only on the suspicious side."

The hypothesis of the conjunctival entry of the trypanosome appears here for the first time, though not discarding the bite.

In the second case he records that (p. 20) "the whole thyroid appears slightly increased in size", adding that (p. 23) "in the facts narrated by the girl's mother, it is the edema of the face and hands which led us to diagnose Chagas' disease."

Romaña does not seem to be conscious that, in this case, there was no unilateral palpebral edema, and that this makes is qualitatively different from the perspective of the sign which he later presents as discovered.

The following article (Romaña 1934b) is of interest since it is the first study of two cases of heart chronic forms published by the *MEPRA*, and also because already in his view (and in that of Mazza's) the chronic illness is fundamentally cardiac (p. 25).

In Mazza & Romaña (1934) they state "what is typical of the clinical symptomatology" (p. 25) which is described (p. 27):

> The first thing that calls our attention in the little patient is the anasarca. The edema is particularly impressive in the eyelids, hands and feet and in the pelvic region, in which the finger does not leave a pit. [...] Eyes: the eyelids appear edematous on both sides, especially in the left eye, which is almost totally closed due to the edema. The palpebral conjunctivas are slightly congested.

Although there is a more intense swelling in one eye, they do not think it is a particularly new sign. We point out that they very clearly speak of edema, not of mixedema.

Mazza makes a very decisive contribution to the article with his finding of a form of trypanosome with two blephoroplasts and two flagella, although with only one nucleus – a sign of the blood division of the parasite –, and with the histological studies of the patient's autopsy. Dr. Jorg, a close collaborator of Mazza's, also contributes with a histological test. The existence of diffuse myocarditis, infiltrative, is verified, of the kind of progressive myocardial fibrosis. (The collective character of the research, which is also observed in the first case of the series, is undeniable.)

The next article in the series (Mazza & Ruchelli 1934, p. 3) shows us a Mazza who is proud of having Romaña as disciple and colleague, and who is also convinced that his findings square with those of Chagas', whom he follows. If the disease is not diagnosed more frequently, it is due to the failure to know the characteristic signs (which are those described by Chagas):

> Some of the collaborators of the Mission (doctor Romaña) as well as one of us (Mazza), have been able to predict clinically the discovery of schizotrypanum in the circulation of certain patients, especially children, with a symptomatic picture which once known, is difficult to mistake for that of other processes, also common in some areas of our country, where we are presently noticing the undreamed of diffusion Chagas' disease has acquired.

Both of the cases described present unilateral bipalpebral edema (left, on this occasion), one of the characteristic signs described by Chagas, and which they find again in their patients, although probably Mazza does not yet suppose that it is the most frequent of the observable signs, that which enables doctors to predict the diagnosis of Chagas' disease "at first sight". He acquires the conviction that this is so only progressively, until he gets to his article about the diagnostic value of "the sign of the eye" in the Ninth Conference of Argentine Regional Pathology, a year later (Mazza 1935b).

We point out that Mazza ascribes the description of the palpebral edema, unilateral or bilateral, to Chagas, before the dispute with Romaña. It is not a resource constructed to deprive Romaña's contribution of originality, as Delaporte holds in his book; an interpretation that leads him to exclaim, *Mazza the fraud.*

In Romaña (1934c), three new cases of Chagas' disease are introduced. In the first of them he finds (p. 21) generalized facial puffiness, in the second (p. 26) "edematous eyelids, pale palpebral conjunctiva."

None of these cases presents Romaña's sign, and show a portal of entry through the skin: one in the temporal region, another one in the groin. The third case finally shows unilateral palpebral edema.

In the previous article Mazza relates this circumstance, considering it one of his contributions to the knowledge of Chagas' disease. He also mentions that the patients present all the signs described by Chagas, except the goiter.

It is remarkable that Romaña expresses the same concepts in his article, almost in the same words, and presents for the first time the terminology that makes such a strong impact on Delaporte, and that refers to "the pure disease".

Let us see how he expresses this (p. 21):

> This fact (he is referring to the absence of endemic goiter and malaria in the north of Santa Fe, which is the scene of his first publication) is very valuable, since it is an area where it was possible to study the *"pure disease"* in several acute cases, enabling the ample ratification of most of the symptoms described by Chagas in his little patients from Lassence, and which he very sagaciously differentiated from the multiplicity of diseases that can be found there.

Again, the signs are those described by Chagas, except those resulting from malaria and the thyroid.

The symptomatology found in the patients is the same Chagas describes (or, at least, that Mazza ascribes to Chagas and teaches to his disciples) and this is clearly stated by Romaña, together with the acknowledgement to his mentor (p. 31): "The last three cases of Chagas' disease we have found in the north of Santa Fe (not having conducted a special search for patients) are described now. Our previous experience enabled us to rapidly suspect the existence of the disease in these new cases, led by the conviction – which professor Mazza created in us – that once the knowledge of the symptomatology of the process is spread among the professionals in the north, the number of patients found will multiply considerably."

In the following publication (No. 21) made by the *MEPRA*, the previous pattern is repeated: an article signed by Mazza in collaboration with Romaña and another author, and a second article signed by Romaña alone. As we shall see, a new factor is introduced here, which will divide the roads between the two authors.

In Mazza, Romaña & Parma (1935), a new case (p. 5) of a two-year-old patient is described, who initially presents

> a little red spot on the left temple. The bite of a *vinchuca* was suspected and the search for this insect led to the discovery of a specimen full of blood under the mattress of the child's bed. The red spot increased in size during the day, and the neighbouring areas became swollen. The following day the edema spread to the left eyelids and continued increasing in size the next day, until the child was unable to open his eye.

Obviously, the portal of entry was the skin, and the edema spread from there to the eye.

The patient dies. The attempts at finding the main author of the article fail once more and show, once more, a collective work (inside the *MEPRA*) in which Bartolomé Parma, who is a doctor at Villa Guillermina the same as Romaña, diagnoses the illness clinically; probably another of the researchers finds trypanosomes in the patient's blood; Romaña does the autopsy, Mazza and Jorg study the histology of the organs; Jorg makes the diagrams which clarify what has been observed in the histological sections. It is interesting to record – in our search for studies about the cardiac components of the disease – that they describe a typical lesion consisting in "progressive cardiac dystrophy, with simple atrophy of the subpericardial myofibrils, proliferation of the reticular connective tissue", and alterations of the spleen with nests of parasites in the spleen. How can they fail to identify the chronic illness with a cardiopathy, if its signs are present even in the acute cases?

It is not here where the controversy begins, but in the article that Romaña signs alone, and which we will present below.

13. The appearance of Romaña's sign as such

It is now, in this paper (Romaña 1935, p. 19) *when Romaña thinks for the first time that what he describes is not present in Chagas' papers*: "The main (because it is totally new) symptom verified in them, unilateral trypanosomal conjunctivitis, will be the object of a special publication, alongside other complementary observations." And it also shows, of course, that Mazza does not strictly impose all his ideas on his collaborators, reason for which the previous acknowledgments to Chagas and to himself undoubtedly come from Romaña himself.

The description he makes of the patients is the following (pp. 3-4):

> About twenty days ago, the left eye began to turn red, as if at the onset of conjunctivitis. In the following days, the eyelids of the same eye became edematous, gradually increasing in size until the palpebral fissure was completely closed. In the morning, on waking up, the eyelids were stuck with some yellowish secretion.

As regards the second case (pp. 26-27)

> he woke up with a swollen and slightly painful right eye [...] the right eye shows eyelids deformed by great edema, which gives the area a slightly purplish colour. This edema spreads onto the neighbouring areas of the face, right cheek and cheekbone, as well as onto the base of the nose.

It is in the next article (Romaña 1935b) when he presents his complete claim: unilateral conjunctivitis is the most frequent initial sign of Chagas' disease and shows the predominant portal of entry of the trypanosome: the conjunctiva. It is a new sign, not previously present in Chagas.

His claim is founded on the cases he studied, since (p. 17)

> up to this moment, we have observed nine acute cases of American trypanosomiasis, and in six of them, that is, in 66% of the cases, we have been able to state clearly that the complaint started with the inflammation of one of the eyes. In these conditions, this fact stops being a mere coincidence in order to become an invaluable symptom to presume the illness at its onset.

The argument on which he supports the hypothesis that the conjunctiva is the main portal of entry is, at least, bizarre (p. 27 and p. 28), because he argues that due to the fact that it is difficult for the feces to get inside the eyes closed during sleep, there is a incongruity between the "high percentage of infected *vinchucas* and the relatively small amount of cases it is possible to identify."

He then comments, in order to discard the skin as the portal of entry that, since (p. 28)

> in certain areas of Argentina 50, 60 or even a higher percentage of the captured specimens, insects which swarm especially during the summer in the huts of our countryside, sucking their inhabitants' blood night after night, it was logical to think that all, or the vast majority of them must have suffered a trypanosomic infection at some moment of their lives, because on being bitten at one moment or another, they must have exposed their skin to the contact of the contaminating feces, that the insect always eliminates while eating.

Romaña is clearly not aware at this moment of the real extent of Chagas' disease, thinking that the quantity of infected patients is not significant. We shall see later that the same premises tend to refute the conclusion about the portal of entry of the disease, if we base our ideas on the epidemiological knowledge that Mazza founds and which is corroborated more and more every day. It is because there are many infected patients that the conjunctiva is not the main entry of the trypanosome.

Delaporte, who is not acquainted with the scientific facts concerning the epidemiological importance of Chagas' disease, reproduces the argument without making any comment, which also Dias (1939b) quotes with signs of approval.

14. The justification of Romaña's sign

There are two questions that need to be asked: the first one is whether the previous observations indeed justify Romaña's assertions that the portal of entry is conjunctival. Romaña says it is, that

> in six cases the infection through one of the eyes (observations 1, 3, 6, 7, 8 and 9) has been clearly demonstrated and in two more (2 and 4), the palpebral edema acquires a predominant role as one of the abnormal manifestations observed by the little patients' relatives. Only in one observation is this detail not visible (case 5).

The second question is why he does not inform us in advance that the sign he describes is new.

Only the revision of the casuistry will enable us to prove the accuracy of Romaña's answer to the first question, since we find it surprising that only just now should he inform us of the finding.

We will refer to the second question later. Its consideration can lead us to an interpretation of the history that – in Romaña – gets nearer to the situation that made Delaporte exclaim that Mazza is a fraud. We will also say that, although the situations are similar, neither one nor the other imply any fraud.

Let us see then, if Romaña's descriptions bear out that the portal of entry of the infection is fundamentally conjunctival (let us remember that in the first article the statement was not so categorical, since there, he expresses that "it could have taken place either through a bite or through *vinchucas*' feces."

If we summarize what was stated in them, we find that two out of the nine cases do not present unilateral palpebral edema; out of the remaining seven, two probably start with bites near the eye (temple or temporal region); in four of the patients, there is no clear conjunctivitis recorded, which is observed in only one of them (let us remember that conjunctivitis is an essential part of Romaña's sign, just like this author describes and Delaporte underlines).

If we reread the list, we find it justified to say that the symptoms mostly start in only one eye, but not that the portal of entry is conjunctival; even on interpreting that the doubtful cases have conjunctivitis, even then, the number of candidates likely to have Romaña's sign *in full* is smaller.

What makes Romaña see as something new what he previously describes as a reassertion of the symptomatology described by Chagas?

We find a first clue in the fact that among his previous articles, in which he doubts about the portal of entry of the disease and these, he is working at the Oswaldo Cruz Institute, where he witnesses an experimental demonstration of it. His testimony leaves no room for doubt (p. 27):

> Indeed, during our stay at the Oswaldo Cruz Institute in Rio de Janeiro last year, we had the opportunity of seeing some cancer patients on whom Evandro Chagas made some experimental infections with American trypanosomiasis. Among them, the only one who contracted the disease was the one infected via the conjunctiva…this patient developed an eye inflammation entirely similar to that observed by us in the acute cases we have referred to, and even the sat-

ellite adenitis completed the picture. However, the attempts at infecting [...] using the epidermis as a portal of entry, were unsuccessful.

And he adds: "This fact would be in agreement with what was already held by Brumpt [...], a claim that must be accepted after the brilliant defense E. Dias made of it."

Let us keep in mind the names of E. Chagas and E. Dias, since they are the protagonists of the episode culminating in the deep enmity between Mazza and Romaña.

If we read correctly what has been said up to now, Romaña (who has known the infection via the conjunctiva since Brumpt, although he does not exclude the infection through bites) becomes convinced, during his Brazilian experience, of the supremacy of the first and re-reads, in this light (retrospectively) the previous material, even though it may not give him all the empirical support which Romaña declares, and needs. It is likewise possible that his new Brazilian friends may have shown him another way of interpreting Chagas' work, in which the edema of only one eye is excluded, contrary to the interpretation he learned with Mazza, where this appears, and with which he makes his first diagnoses. For this reason, perhaps, Dias (1936, p. 345) expresses that "There is a curious circumstance that must be mentioned, and it is that that sign went unnoticed in the eyes of the researchers who were studying the disease in Brazil." The cycle has been completed and Romaña informs us, in all good faith, that these diagnoses are a novelty from the beginning, that he has always seen them this way.

This is the answer to the second question. In the first seven cases, Romaña does not talk about a novelty, because at that moment, he does not think it is one. Only after August 1934, when he witnesses E. Chagas' experiments and even later, does he begin to consider his findings as original.

Romaña's works and, above all, the reinterpretation he makes of the symptomatology of the cases he published up to that moment and which lead him to his strong claim of the way of infection via the conjunctiva, enable Brazilian researchers to add the clinical signs of the unilateral palpebral edema and the accompanying adenopathy (which, according to them, have not been observed in Brazil) to their laboratory experiments and assign real significance to their laboratory findings.

As Mazza sharply points out years later (c. 1940, p. 22) right in the middle of the dispute, E. Dias replaces, referring continuously to Romaña's schizotrypanosic conjunctivitis, the "lack of observation material of schizotrypanosis of their own in Brazil where, nevertheless, it must be frequent."

How does Mazza react to these assertions?

Let us remember once more that Romaña's articles are disseminated by Mazza through the Review of the Mission. He not only agrees – though only partially – with Romaña. He supports him completely. An article published barely two months later (Mazza & Govi 1935, p. 19) is a curious indicator (if we see it in the light of the following disputes) of this. In it he expresses: "the symptomatology clearly described for the acute cases of Chagas' disease, mostly the unilateral schizotrypanosic conjunctivitis accompanied by fever and a bad general condition, especially in children, defined lately by Romaña, enabled one of us (Govi), while being in the presence of a patient with that clinical picture, to make a diagnosis immediately."

This article is dated 8th. June, 1935. In October 1935, on the occasion of the Ninth Meeting of the Society of Regional Pathology which takes place in Mendoza, Mazza (1935b) insists "on the value of the palpebral edema on one side of the face in the diagnosis of the acute stage of Chagas' disease."

In that same meeting, something happens, which years later Mazza sees as the beginning of his differences with Romaña (Mazza, circa 1940):

> To this congress, held in memory of Carlos Chagas, who died last year, came Evandro Chagas, where invited his son and E. Dias, as delegates from the Oswaldo Cruz Institute. Despite the presentation of the two cases from San Juan we are dealing with, of those already well-known in Argentina and of another thirty-three, presented only in the course of the meeting, which showed only exceptionally the existence of "schizotrypanosic conjunctivitis", the above mentioned doctors, with an unknown purpose in mind but evidently thought over in advance, without having contributed with casuistry of their own, proposed the designation of the ophthalmic symptom as the sign of one of the doctors who had followed our inspiration and in-

structions, applying the knowledge acquired through the fundamental teachings of Carlos Chagas.

Mazza is surprised at the Brazilian initiative, but he does not drift apart from Romaña, with whom he continues collaborating and together with whom he publishes a new article the following year (Mazza, Romaña & Parma 1936), where "the hypothesis of the entry of the *Schizotrypanosum cruzi* through the skin becomes confirmed".

It is necessary to stress two facts here. The first one is that in this article, Romaña's argument that the difficulty of the conjunctival transmission during sleep accounts for the small number of patients, is inverted; in this case, it is thought that the skin portal of entry justifies the great expansion of Chagas' disease, due to the continual bites of *vinchucas*. It is in this argument perhaps, that Mazza's obstinate opposition to thinking that the disease enters – mostly – through the conjunctiva, lies; in the belief, fully confirmed later, that the disease is of crucial epidemiological importance.

The second point we want to stress is even more obvious. Romaña still thinks that the bite is a good alternative against conjunctival transmission, as he expresses it in his first article.

This is not the last article Mazza and Romaña do together. There is an almost immediate case of skin penetration detected by Zambra, also in Villa Guillermina, and studied by both authors (Mazza, Romaña & Zambra 1936).

15. The estrangement

Approximately at that moment the manifest enmity between them starts, which coincides with the publication of E. Dias' writings (1936, 1939a, 1939b). In them, apart from attributing the discovery of a pathognomonic sign to Romaña, he expresses that, due to this fact, the cases found by the *MEPRA* (i.e. by Mazza and his collaborators) and on which the revival of the scientific interest in Chagas' disease is based, were discovered thanks to the fact that the patients manifested Romaña's sign.

Mazza's work, then, ceases to have value *per se* and becomes subsidiary to Romaña's findings.

Delaporte rightly says that Mazza is faced with the insufferable. He is not the one that revives Chagas' disease. His work is secondary.

What is important is that one of his collaborators, having studied nine cases – in which he himself took part – highlights a group of symptoms he thinks are already present in Chagas, in order to get all the merit.

However, despite his sharp remarks, Mazza behaves in a context of discussion which is essentially scientific as well as historical. He puts great effort, perhaps until his death, into demonstrating the first, because he understands that to attribute the so-called Romaña's sign to all the acute patients, hides the signs of the most severe forms of the disease from the ordinary doctor's perception, those which end up in death, and whose patients could be saved if the medicine 7602 Bayer was used in time (Mazza is the first one to prove this medicine to be effective on the illness).

It is not that he denies the importance of the identification of the unilateral edema of the eyes in order to suspect Chagas' disease, mostly if it is accompanied by other signs such as fever, tachycardia, weakness, etc.

From the scientific point of view, Mazza holds that the portal of entry, in a large percentage of cases, is the skin, and not the conjunctiva. In this context, the unilateral palpebral edema is secondary to the bite, which usually takes place in the face, near the eyes, since it is the part of the body that is not covered during sleep. And the conjunctival reaction – if there is one – is also secondary.

From the historical point of view, he states that the unilateral palpebral edema is not a discovery, since it had already been described by Carlos Chagas.

It is not our intention to solve a scientific discussion on the basis of the writings of that time (a historian does not solve a scientific problem; this is done by the scientific community); our intention is to lay the foundations in order to understand the points of view involved, beyond the intention of its protagonists, who have never been regarded as irrational, even in the midst of violent discussions or personal resentment. We shall later see the present points of view on Chagas' disease, and how the scientific community solved – Solomonically – the dispute between both researchers.

As regards the historical divergences, we know that the question of priority is one of the nastiest and most awkward problems since, as we have already said, it involves questions of fact and conceptual ones, so that, instead of being faced with an event in particular, we are faced with a process in which it is difficult to assign priorities. Probably the solution lies in thinking that scientific knowledge is a collective construction with unequal contributions to objectives shared by a group of historical agents, with no heroes who bear upon themselves all of the burden.

It is probably this historiographic perspective, which separates us more distinctly from Delaporte. Where Delaporte finds Romaña's privileged role, we see the meeting of a community of researchers who support one another and in which Mazza's role is essential. Where Delaporte sees Romaña going, unfalteringly, along a new path, we see him going along a path *with others*, and precisely thanks to the fact that there *are* others; that path has comings and goings that resignify what came before, and also shows as new what was not, and in which this novelty is constructed as such in the midst of simplifications that end-up legitimating it later.

16. The scientific problem

Delaporte sees the union between a clinical symptomatology and a way of entry as one of the main scientific contributions of Romaña's. If Romaña refounds Chagas' disease, the portal of entry, verified by the Brazilian scientists and by Romaña himself, is the one that characterizes the disease.

Mazza does not discuss the existence of the ofthalmoganglionar complex (of the sight and the ganglia nearest to the eye) as a sign of the acute illness, but he believes – also after some comings and goings – that the conjunctivitis that comes along with it, does not always indicate the portal of entry, but is secondary to the entry of the parasite through some skin lesion – bite or scratch – next to the eye, the same as the edema and the ganglia.

Years later, Mazza (circa 1940) sets the change of opinion around the end of 1935, since "in the course of that year and the following ones, and after reconsidering the observations published, we had to conclude the non-existence of 'conjunctivitis' and the presence of only an eritrochromial edema of eyelids, with greater or lesser secondary conjunctival reaction."

Our own revision of the cases tends to weaken Romaña's (later) assertion about the supremacy of the conjunctival portal of entry, even in his own observations, which show, in his own commentaries, other portals of entry.

The joint publication made by Romaña and Mazza of two cases in which the portal of entry was the skin, persuades us that Mazza had, in fact, already abandoned his previous enthusiasm for the conjunctival way of transmission, and of Romaña's persistence of a much more shaded point of view than that Delaporte presents.

But this does not mean that he discards studying the role of the conjunctiva in Chagas' disease. On a date as late as 1937, the *MEPRA* publishes an article (Olle 1937) in which an ophthalmologist from Santiago del Estero studies – at Mazza's request – (Mazza does most of the patients' blood tests) the possibility that those patients who come to consult an ophthalmologist and present unilateral edema of the eyelids and regional ganglia, might be parasitized by the *Trypanosoma cruzi*. In the ten observations he relates and which refer to patients examined between April and August 1936, there is proof of the presence of the parasite in the blood; three of the patients do not show any sign of conjunctivitis.

In 1936 Mazza studies the biopsies of the conjunctiva of Chagasic patients with acute ophthalmoganglionar symptoms, and publishes – together with Jorg – his findings at the Sixth National Congress of Medicine in Cordoba in October 1938, in Volume III of its minutes, appearing in 1939. From the fifty studies made, it was inferred that the conjunctival inflammatory reaction was secondary, caused by the spreading of the deep inflammatory state originated in Tenon's capsule or in the cellular tissue of the orbit. They never found any parasites inside the epithelial cells of the conjunctiva, either in the histological sections or in smears, though they did find them in the histocytes of the area, since the parasite always ends up in histiocytic macrophagia. He concludes that up to that moment, nobody has demonstrated epitheliotropism (i.e. the tendency to head towards the epithelia, skin or conjunctiva) in the *Trypanosoma cruzi*. From these results, Mazza can validly argue that the penetration of the parasites into the epithelial cells of the conjunctiva, and from there into the histiocytes and into the blood, is relatively difficult. A full mark in support of his claim that the conjunctival inflammation is secondary and not primary.

For his part, Romaña (1939a, 1939b) attempts – successfully – to transmit Chagas' disease to a monkey via the conjunctiva, causing a strong infection in it, which affects the conjunctiva and the periocular tissues, with all the characteristics of the human ophthalmoganglionar complex, and in which he verifies the trypanosomic invasion of the epithelial cells.

Mazza's reply is (1939a, 1949) that, in the first place, the infection in laboratory animals could be different from that in human beings; in the latter, the conjunctival reaction is slight and, as it happens in the vast majority of cases, even in Romaña's, in which there is no secretion – which, for Mazza (and not only for him) is the effective indicator of conjunctivitis –. On the other hand, he interprets that as the eye symptoms of the illness appear only twelve days after the conjunctival inoculation – and only after ten days are trypanosomes found in the blood –, they are secondary lesions, transmitted to the eye via the blood.

As we can see, the arguments on both sides are of a scientific nature and are both praiseworthy, which makes it difficult for a historian to take sides with either of the two actors. In Mazza's case, they are rather marred because they are mixed with expressions of anger, but are nevertheless valuable.

It is perhaps for this reason that Delaporte does not notice them and listens only to Romaña and his allies, the Brazilian researchers E. Chagas and E. Dias. However, we cannot fail to point out that when he takes sides, he makes unforgivable mistakes. The one we want to stress at this moment is that which assigns superiority to Romaña because his scientific position is the correct one.

In the vivid description Delaporte makes of the lecture Romaña gives in the presence of Carlos Chagas, he indicates that (p. 159) "in the twilight of his life, Chagas certainly becomes aware that the essential was missing in his claim: the palpebral edema, linked to the conjunctival infection." He adds (p. 160): "From the moment Romaña starts thinking that Chagas' disease is a parasitosis close to other trypanosomiases, the palpebral edema shows the portal of entry", and concludes (p. 166)

> Romaña thought he could dispel what was still a paradox at that moment: the contrast between the great dissemination of the infected insects and the small number of cases recorded up to then. We mentioned how Chagas solved this difficulty: with the frequency of the bites which imply the inoculation of the parasite, there is a general infection and it goes unnoticed. From the moment in which the edema becomes the sign of the infection, the infection is no longer general. The substitution of the inoculation portal of entry for the contamination portal of entry leads Romaña to link the bite of the insect to the risk of infection. All individuals are certainly exposed to the contact with the contaminating feces that the insects eliminate when they feed. But the risk of contracting the disease is small, if the mode of contamination is taken into account. The edema shows that the natural portal of entry is the eye mucosa. Although it is inaccessible during sleep, it is nevertheless a restriction factor.

Both the discovery of the ophthalmoganglionar complex, together with the conjunctival portal of entry – which is the natural portal of entry – and Romaña's explanation of the few clinical cases found (precisely because it is difficult to contract the illness in this way), are the great contributions that refound epistemologically – according to Delaporte – Chagas' disease.

Later we will refer to the third factor which appeals to the condition of possibility of these discoveries, caused by the separation Romaña makes between the parasitosis and an endocrine illness, as Carlos Chagas conceived it, and to the discovery or rediscovery of the unilateral palpebral edema.

For the time being, in the context of the discussion between Mazza and Romaña, we simply point out that the conjunctiva is not the main portal of entry, and that the illness becomes widespread precisely because it is not.

As Delaporte does not consult present-day texts about Chagas' disease – he sticks exclusively to the material contemporary with the discussion – he takes as accepted scientific knowledge what E. Dias says. Nevertheless, this is not so, as we shall see later.

The present consensus is that the conjunctiva is not the privileged portal of entry, and that Mazza's points of view are widely accepted.

As the World Health Organization says (WHO 1955, pp. 125-126):

> The triatome feeds during the night and, as it is attracted by the exhalation of carbon dioxide, it falls on the sleepers' beds in order to feed on their exposed skin, very often on their face near the mouth. Nevertheless, the parasites are not transmitted through the insect's bite. They are deposited on the victim's skin with the insect's feces. *When the victim scratches the bite, the parasites are inadvertently helped to penetrate the skin and enter the bloodstream* (italics by C. L.).

In the abovementioned paper, which is central to the area of public health nowadays, and where there are a few paragraphs which refer to the biomedical aspects of the disease, the privileged portal of entry is the skin whereas the conjunctival portal of entry is not mentioned.

In a well-known medical text (Isselbacher 1994b, Vol. 1, p. 1044), where all the aspects of the disease are dealt with in depth, the portal of entry through the conjunctiva is ranked third, after the skin and the mucosae.

The only illustration of a patient with acute Chagas' disease found in both texts is the photograph of a child with unilateral palpebral edema and referring to it as "Romaña's sign".

This is because – as Mazza holds – different portals of entry manifest themselves secondarily as an ophthalmoganglionar complex. It constitutes the most characteristic image of Chagas' disease, although it does not represent an ophthalmic portal of entry.

Even so, it is not necessary to get to the present time to know what the scientific community thinks about these questions. A year after Romaña's *Replica* to Mazza's objections and when passions calmed down, he publishes a twelve-page dissemination booklet in which he explains to the general public what Chagas' disease consists of. In it he expresses (p. 6):

> The only way of transmission of Chagas' disease is through poop or *vinchucas'* feces. The parasites enter through the skin or the mucosae, in the area close to the bite and produce a local inflammation five to ten days after, which country people commonly call "mal de vista" (disease of the sight) and also "aire"(air). If, for example, this local inflammation is on the skin of the arms or legs, it looks very much like a boil, and if it is near the eye, it looks like conjunctivitis.

As we can see, the conjunctiva is not the privileged portal of entry of the disease and the bite produces – generally, though not always – the unilateral palpebral edema, but only when it is close to the eye.

Delaporte, who reads one of Romaña's articles of the year 1944 – *Replica* – , but not one of the year 1945, repeats almost sixty-five years later that his sign is pathognomonic and that it indicates that the portal of entry is conjunctival. Scientific mistakes that also Dias makes – although he does not go to the extreme of thinking that the sign is pathognomonic (1939, p. 967: "the importance of the symptom was such that, even without being strictly pathognomonic") – and which Romaña adheres to for a short time.

Let us consider now his assertion that Romaña makes a radical epistemological change when he separates Chagas' disease from the endocrine diseases.

Indeed, Romaña studies Chagas' disease in an area in which there is no endemic goiter, something he probably does at the request of Mazza. But this does not mean that they had separated it completely from the thyroid disturbances. There is a possibility that they deferred reaching a decision, and explored all the possibilities, all the hypotheses. After all, when Niño (1929) summarizes the eighteen observations of American trypanosomiasis found in Argentina, the great majority of them show goiter.

As we can see, the confusing association between goiter and trypanosome that misleads Chagas is likewise observed in Argentina.

This dual attitude is best shown in one of Romaña's articles (1935c), which presents a synthesis of the discussion about the link between American trypanosomiasis and goiter. He starts by saying that (p. 897) "I want to establish that, on dealing with this subject, it is not my purpose to arrive at definite conclusions on such a debated matter", and continues presenting the arguments in favor of and against the theory of the Chagasic origin of goiter, six reasons in favour of it and five against it.

The article finishes without Romaña making any statement on the question at a moment in which he, according to Delaporte, had made an epistemological revolution when he separated both diseases by means of the sign that bears his name.

Once more we verify that the construction of knowledge is social, not individual, and it is carried out by moving along roads which are full of contradictions and persistent unpleasant memories of the past, instead of by the sharp cuts that Delaporte seeks, but does not find.

17. The roles in history

Let us remember his terms once more. According to Delaporte, Romaña discovers the sign that bears his name, and which is an oculoganglionar symptomatic complex consisting in unilateral conjunctivitis, accompanied by edema of the eyelids and regional adenopathy. In Mazza's opinion, the unilateral edema of the eyelids is present in Chagas' writings. Consequently, Romaña does not discover it. It is Mazza, who follows Chagas' teachings, who shows the sign to Romaña, alongside whom he makes his first diagnoses.

We have basically seen that this is so. When we go through the pages of Chagas' 1916 article, we verify that in various pictures and in some descriptions, the patient shows unilateral edema of the eyelids. In them, the same as in the photographs included in Mazza's article and in present-day texts about Chagas' disease, we see Romaña's sign.

We can understand that Mazza sees in the pictures the same manifestation we see: unilateral edema of the eyelids, and he passes it on to all those who listen to him talking about Chagas' disease. But he does not separate it from the rest of the signs, but rather as Chagas' himself presents it, together with the swollen face, both eyes edematized, etc. Dias says (1939b, p. 969) that the facies described by Chagas "cannot be confused with the more or less localized palpebral edemas and which are accompanied by other signs." This is not true, because we *have* confused them. Niño, who is a faithful exponent of Chagas' thoughts (in the reading of the Argentine researchers formed by Mazza), expresses that (1929, p. 152) "the clinical picture of the acute form is characterized by a series of symptoms that bear the pathognomonic hallmark: mucous degeneration of the subcutaneous tissue, giving the edema a special character: edema, preferably located in the eyelids."

That is why any of them diagnoses the acute stage of Chagas' disease correctly when they are faced with a patient with unilateral edema of the eyelids, even though they do not know that it is so frequent or so meaningful as they learn later.

Romaña saw this, although the edema is not pathognomonic, although the conjunctiva is not its privileged portal of entry (in Delaporte's point of view, the only one), although it is not accompanied by conjunctival suppuration or in spite of his not having categorically excluded thyroid lesions.

This is why the symptomatic complex of the unilateral bipalpebral edema, accompanied by satellite adenopathy and conjunctival irritation rightly bears his name and appears under the photographs of patients suffering from acute Chagas' disease and appearing in all the publications dealing with the subject.

As far as Mazza is concerned, he was not a fraud. He was one of the founders of Argentinian parasitology, the person who studied and spread all the aspects of Chagas' disease, not leaving anything out of his insatiable curiosity, verifying or refuting even the most minute detail. He probably died of the same illness he fought against with all his might. His proverbial bad temper makes him look down on Romaña. But this is an ad-hominem argument that cannot dim his scientific work. His name is deservingly linked to that of Carlos Chagas' every time American trypanosomiasis is mentioned.

When everyone calms down, Romaña (1958) summarizes the history and the characteristics of Chagas' disease, puts the contributions of his mentor Salvador Mazza, in due historical perspective (p. 190):

> Chagas managed to prove, at a famous public trial, the fundamental truth of the clinical picture he had described, but he lacked data to confirm his epidemiological conception, even from Brazil itself. Only years later could his fundamental ideas begin to be recognized as true. The reaction started in Argentina with Professor Mazza at the head of a group of inland doctors whom he had taught to discover the reality of the pathological world that surrounded them. In 1934 I had the honour to take the dawn of this truth to the erudite Academia de Medicina in Rio de Janeiro and where Chagas then, in the twilight of his life, had the opportunity to hear it; there, where he had been strongly fought against.

Romaña's great human kindness goes beyond past antagonisms and refers to Mazza as what he is, the person responsible for the revival of Carlos Chagas' studies.

Romaña's sign is, at present, universally acknowledged, and we find their divergences difficult to understand, beyond personal differences, in which Mazza's characteristic bad temper must have played an important role.

In spite of Delaporte, Romaña is not the hero he depicts. Neither is Mazza a villain. History is not made by heroes or villains but simply by men, with their virtues and flaws, who are right and wrong, but who build with others that structure of thought that exceeds them, which is scientific knowledge. And this does not always present – at least in biological disciplines – those sharp divisions which, in other sciences, separate the successive stages of knowledge. On the contrary: its evolution is much more similar to biological evolution than one would think, where small successive changes lead to the transformation of what is known.

We had questioned Delaporte's historiographic approach, because it does not allow him to recognize the collective construction of scientific knowledge, which feeds from multiple contributions, in which a careful account hardly ever allows us to isolate the exact moment in which the novelty appears or the individual who discovers it. But what is more serious is his methodological restriction to the writings of that time, because it makes him commit scientific errors.

For this reason, Delaporte's Chagas is brilliant, well-informed, always provocative, but with flaws which are necessary to correct if we intend to understand with the greatest precision that outstanding period of the Latin American history of science and the role that the men who constructed the present knowledge of American trypanosomiasis played, in spite of all the obstacles.

References

Aguirre, J.A. & C. Giménez (1938), "Consideraciones de semiología radiológica sobre 168 roentgencardiometrías en la enfermedad de Chagas", *6° Congreso Nacional de Medicina*, Vol. 3.

Brumpt, E. (1912), "Pénetration du Schizotrypanum cruzi a travers la muqueuse oculaire saine", *Bulletin de la Societé de pathologie exotique* 5: 22-26.

Brumpt, E. (1913), *Précis de parasitogie*, Paris: Masson.

Bennett J.C. & F. Plum (eds.) (1996), *Cecil Textbook of Medicine*, 20th ed., Philadelphia: W.B. Sanders.

Chagas, C. (1941) "Tripanosomiasis Americana. Forma aguda de la enfermedad", *Misión de Estudios de Patología Regional Argentina (MEPRA)*, *Publicación* 55: 3-45.

Chagas, E. (1933), "Infection expérimentale de l'homme par le Trypanosoma cruzi", *Comptes rendus de la Société de biologie* 115: 1339-1340.

Chagas, E. (1934), "Infection expérimentale de l'homme par le Trypanosoma cruzi", *Comptes rendus de la Société de biologie* 117: 390-392.

Chagas, E. (1935), "Infection expérimentale par le Trypanosoma cruzi chez l'homme", *Comptes rendus de la Société de biologie* 118: 718.

Crowel, B.C. (1923), "The Acute Form of American trypanosomiasis: Notes on Its Pathology, With Autopsy Report and Observations on Trypanosoma cruzi in Animal", *American Journal of Tropical Medicine* 3: 425-454.

Delaporte, F. (1999), *La maladie de Chagas*, Paris: Payot.

Dias, E. (1936), "O Signal de Romaña e os novos progressos no estudo da doença de Chagas", *Folha Médica* 17: 345.

Dias, E. (1939a), "O signal de Romaña na molestia de Chagas", *Acta Médica* 3 (4): 60-62.

Dias, E. (1939b), "O signal de Romaña e sua influencia na evoluçao dos conhecimentos sobre a molestia de Chagas", *Brasil-Médico* 53 (42): 5-10.

Dias, J.C.P (1979), "Epidemiological Aspects of Chagas' Disease in the Western of Minas Gerais", en *Congresso Internacional sobre Doença de Chagas*, *Abstracts*, Rio de Janeiro: [s.n.], pp. 1-6.

Guerreiro Cezar, M.A. (1913), "Da reacção de Bordet y Gengou na molestia de Carlos Chagas como elemento diagnóstico", *Brazil-Médico* 27: 225-226.

Isselbacher, K., Braunwald, E., Wilson, J., Martin, J., Fauci, A. & D. Kasper (1994), *Harrison. Principios de Medicina Interna*, Buenos Aires: Interamericana-McGraw-Hill.

Krauss R., Maggio C. & F. Rosenbush (1915), "Bocio, cretinismo y enfermedad de Chagas (1ª Comunicación)", *La Prensa Médica Argentina* 1: 2-5.

Krauss R. & F. Rosenbush (1916), "Bocio, cretinismo y enfermedad de Chagas (2ª Comunicación)", *La Prensa Médica Argentina* 17: 177-180.

Magarinos Torres, C. (1935), "Patogenia de la miocarditis crónica en la enfermedad de Chagas", *Novena Reunión de la Sociedad de Patología Regional, Mendoza*, Buenos Aires: Imprenta de la Universidad, pp. 902-916.

Mazza, S. (1935a), "Forma crónica cardíaca de la enfermedad de Chagas comprobada por inoculación en el Departamento El Carmen, Jujuy", *Novena Reunión de la Sociedad de Patología Regional*, Mendoza, p. 418.

Mazza, S. (1935b), "Sobre el valor del edema palpebral de un solo lado para el diagnóstico de la forma aguda de la enfermedad de Chagas", *Novena Reunión de la Sociedad Argentina de Patología Regional, Mendoza*, Buenos Aires: Imprenta de la Universidad, pp. 343-345.

Mazza, S. (1939a), "Inexistencia de un síntoma patognomónico en formas agudas de enfermedad de Chagas", *La Prensa Médica Argentina* 38: 1569-1579.

Mazza, S. (1939b), "Método de investigación de la epidemiología de la Enfermedad de Chagas. La viscerotomía cardio-hepática", *La Prensa Médica Argentina* 50: 2461-2470.

Mazza, S. (1940), "Enfermedad de Chagas en San Juan. Consideraciones generales", *Misión de Estudios de Patología Regional Argentina (MEPRA), Publicación* 43, B: 20-35.

Mazza, S. & C. Benítez (1937), "Comprobación de la naturaleza esquizotripanósica y frecuencia de la dacrioadenitis en la enfermedad de Chagas", *Misión de Estudios de Patología Regional Argentina (MEPRA), Publicación* 31: 3-31.

Mazza, S. & R. Olle (1936), "Particularidades de dos casos de enfermedad de Chagas", *Misión de Estudios de Patología Regional Argentina (MEPRA), Publicación* 28 (1): 3-12.

Mazza, S. & C. Romaña (1931a), "Nuevas observaciones sobre la infección de armadillos del país por el Tripanosoma cruzi", *La Prensa Médica Argentina*, February 28, 1931.

Mazza, S. & C. Romaña (1931b), "Infección espontánea de la comadreja del Chaco Santafecino por el Tripanosoma cruzi", Ponencia en *Séptima Reunión de la Sociedad Argentina de Patología Regional del Norte*, Tucumán, October 5, 6 and 7, 1931.

Mazza, S. & C. Romaña (1933), "Comprobación de Panstrongylus (Triatoma) geniculatus, vinchuca de los tatús, en el norte santafesino", Ponencia en *VIII Reunión de la Sociedad Argentina de Patología Regional del Norte*, Santiago del Estero, October 3, 1933.

Mazza, S. & C. Romaña (1934) "Otro caso de forma aguda de enfermedad de Chagas observado en el norte santafesino", *Misión de Estudios de Patología Regional Argentina (MEPRA), Publicación* 15 (2): 25-54.

Mazza, S. & C. Romaña (1935), "Nota complementaria para la publicación No. 15, II sobre un caso de forma aguda mortal de enfermedad de Chagas en el norte santafecino", *Misión de Estudios de Patología Regional Argentina (MEPRA), Publicación* 2: 19-21.

Mazza, S. & A. Ruchelli (1934), "Comprobación de dos casos agudos de enfermedad de Chagas en Tinogasta (Catamarca), *Misión de Estudios de Patología Regional Argentina (MEPRA), Publicación* 20 (1): 3-19.

Mazza, S., Romaña, C. & B. Parma (1935), "Un nuevo caso mortal de enfermedad de Chagas observado en el norte santafesino", *Misión de Estudios de Patología Regional Argentina (MEPRA), Publicación* 21 (1): 3-19.

Mazza, S., Romaña, C. & B. Parma (1936), "Caso agudo de enfermedad de Chagas con lesión cutánea de inoculación", *Misión de Estudios de Patología Regional Argentina (MEPRA), Publicación* 28 (4): 29-33.

Mazza, S., Romaña, C. & E.R. Zamba (1936), "Comprobación de la lesión cutánea de inoculación en un caso de enfermedad de Chagas", *Misión de Estudios de Patología Regional Argentina (MEPRA), Publicación* 28 (5): 34-40.

Mazza, S., Montaña, A., Benítez, C. & E. Janzi (1936), "Transmisión del Schizotrypanum cruzi, al niño por leche de la madre con enfermedad de Chagas", *Misión de Estudios de Patología Regional Argentina (MEPRA), Publicación* 28 (6): 41.

Niño, F. (1929), *Contribución al estudio de la enfermedad de Chagas o Tripanosomiasis americana en la República Argentina*, Tesis de Doctorado, Facultad de Medicina, Universidad de Buenos Aires.

Niño, F. (1936), *Contribución al estudio de la distribución geográfica de enfermedad de Chagas comparada con la de los triatomas vectores del Schizotripanosoma cruzi en la República Argentina*, Trabajo de adscripción a la cátedra de parasitología de la Facultad de Medicina de la Universidad de Buenos Aires, typescript.

Olle, R. (1937), "Síntomas oculares de la enfermedad de Chagas. Su significación diagnóstica", *Misión de Estudios de Patología Regional Argentina (MEPRA)*, *Publicación* 30 (3): 30-49.

Romaña, C. (1931), "Infección espontánea y la experimental del tatú del Chaco Santafecino por el Tripanosoma cruzi", *Séptima Reunión de la Sociedad Argentina de Patología Regional del Norte*, Tucumán, p. 969.

Romaña, C. (1934a), "Comprobación de formas agudas de tripanosomiasis americana en el Chaco austral y santafecino", *Misión de Estudios de Patología Regional Argentina (MEPRA)*, *Publicación* 1: 3-24.

Romaña, C. (1934b), "Comprobación de formas crónicas cardíacas de tripanosomiasis americana en el norte santafecino", *Misión de Estudios de Patología Regional Argentina (MEPRA)*, *Publicación* 2: 25.

Romaña, C. (1934c), "Nuevas comprobaciones de formas agudas puras de enfermedad de Chagas en el norte santafecino", *Misión de Estudios de Patología Regional Argentina (MEPRA)*, *Publicación* 20 (2): 19-31.

Romaña, C. (1935a), "Dos casos agudos más de enfermedad de Chagas en el norte santafesino", *Misión de Estudios de Patología Regional Argentina (MEPRA)*, *Publicación* 21 (2): 20-32.

Romaña, C. (1935b), "Acerca de un síntoma inicial de valor para el diagnóstico de forma aguda de la enfermedad de Chagas. La conjuntivitis esquizotrypanósica unilateral. (Hipótesis sobre puerta de entrada conjuntival de la enfermedad)", *Misión de Estudios de Patología Regional Argentina (MEPRA)*, *Publicación* 22: 16-28.

Romaña, C. (1935c) "Tripanosomiasis americana y bocio endémico. Estado actual de la cuestión", *La Semana Médica*: 897-902.

Romaña, C. (1939a), "Reproduction chez le singe de la conjontivite schizotrypanosomienne unilatérale", *Bulletin de la Société de pathologie exotique* 32: 390-394.

Romaña, C. (1939b), "Le parasitisme des cellules épithéliales de la conjonctivite du singe", *Bulletin de la Société de pathologie exotique* 32: 810-813.

Romaña, C. (1945), *Qué es la enfermedad de Chagas*, Instituto de Medicina Regional, Universidad Nacional de Tucumán, Publicación 382.

Romaña, C. (1958), "La enfermedad de Chagas, problema social americano. Cómo resolverlo", *Anales de sanidad* 1 (3-4): 189-198.

Romaña, C. (1963), *Enfermedad de Chagas*, Buenos Aires: López Libreros.

World Health Organization (WHO) (1995), "Chagas' Disease", *Tropical Disease Research. Progress 1974-94. Highlights 1993-94*, Geneva: TDR, pp. 125-133.

A Brief Reading of the Iatrochemistry in R. Bostocke: A Philosophical Summary of Man's / The Universe's

Ivoni Reis

Catholic University of São Paulo (PUC-SP)

Since the beginning of human times, the man has often been visit by unknown diseases, along with the already existing ones, which are all frequently difficult to treat.

The physicians are forced to find help to face up to these situations within their knowledge of the subjects during their period of study. Obviously philosophies and religious beliefs are added to the "scientific" knowledge in this unending search.

These attempts often seem to go round in circles like windmills. Truths about health care, which today are backed up by nutritionists, endocrinologists, and by the most modern and the most daring SPAS are found in quotations by Hippocrates[1] in the fifth Century before our era.[2]

---

[1] Better saying, in the *corpus hippocraticum* – a set of writings attributed to Hippocrates and his disciples, who started to have meetings in the Alexandria Library, as of the 3rd century AC. "Contemporary physicians of the Hippocratic school already made little use of drugs, relying on diets and the *vis naturae medicatrix* to achieve the cure", Jones, "Introduction" to Pliny (1989, Vol. VI, pp. xvi-xvii).

[2] Bostocke mentions in his book, that "They [Machaon and Podalírio, sons to Esculápio] thought that, while man made use of a good diet, did exercises and had a

One of these attempts, the humoral theory of sickness, deserves to be pointed out in the brief panoramic history we are presenting in this work. This theory joined with those of the four elements has prevailed for fourteen centuries, leaving traces, which can still be seen, in popular medicine.

In Century II, the humoralist medicine took a big jump with Gallen of Pergamo, who would change the hypocritical humoralist as well as the ancient theories about the four elements into one new medical theory.

Over the centuries, different currents of Gallenism were developing. Perhaps the most important one was iatrochemistry.

Even though the origin of iatrochemistry, or medical chemistry, got lost in time, one of the names to honor them since medieval times was that of the Persian Physician Abu Bakr Muhannad ibn Zacariyya'al-Razi, better known by his Latin name, Razes (865-923). His book *Kitab al-Mansuri*[3] was, in a way, obligatory reading for all "scholars" in the middle Ages till the end of the Sixteenth Century.

In one of his many books, *Doubts About Galen,* Razes comments "Medicine is a type of philosophy and is therefore not compatible with resigning to the criticisms related to the first authorities..." (Razes, *apud.* Alfonso-Goldfarb 1999, p. 37).

In an attempt to broaden and update his medical knowledge at that time, Razes worked with the Arab system of sulphur-mercury and ended up introducing salt (Alfonso-Goldfarb 1999, p. 36), which becomes known later as the Paracelsus triad.[4]

Upon approaching medical alchemy, Razes finally suggests the use of mineral elixirs as medicines (Razes, *apud.* Alfonso-Goldfarb 1999, p. 37).

His main interest was always to take what he considered useful and true from the ancients, despite always reminding us that these teachings should be adapted to the demands or each one's times and

---

good life organization, they would be healthy and would prolong their lives smoothly", Bostocke (1585, sig. G.iif.(r)).

[3] Which in 1170, was translated to Latin by Gerardo de Cremona, called *Liber de Medicina ad Almansorem.*

[4] Subject that we talked about a little, throughout this work.

to each one's own experience (Pagel 1982, pp. 252-253), we quote the physician from Valencia – Arnald of Villanueva (1235-1311).

One of the ideas, which are present in his work, is the bodily concepts and the spirituality ones, which gave him the chance to introduce the *"Quinta Essentia"*[5] concept.

According to Walter Pagel, a history of medicine and the iatrochemistry Scholar, some of Arnald of Villanueva's theories preceded Paracelsus', such as his appraisal of each disease as specific entities, resulting from a specific cosmic constellation. However, still according to W. Pagel, Arnald's concern with alchemy, in no way altered his adherence to humoralism (Pagel 1982, p. 258), and it is specifically at this point that Arnald's medical philosophy differs from that of Paracelsus.

With the discovery of "new worlds", new sicknesses and new medicines invaded Europe, at which point it was natural that the interest in new researches and new ways of healing became of utmost importance. Many epidemics resisted on the methods used by humoralist medicine, which continued to be considered terrible in European academic circles. The search for knowledge, sometimes much more ancient than others totally unknown drove these scholars to put more hope into Razes' elixirs and with Arnald of Vilanueva's quintessences, thus new chemical medicines and ancient elixirs were constantly present in medical prescriptions during the whole of the 16th Century.

In England, the first trustworthy writings about medication, and distilled remedies, appeared via Conrad Gesner's famous hands with the worthy contribution of names like Hieronimus Brunschwig and Arnald of Villanueva (Debus 1996, p. 52).[6] This fact lead to the British accepting chemically prepared medicines better than the European continent. Nonetheless, one cannot say the same about chemical philosophy as a whole.

---

[5] The spirit of the earthly things, captured and transformed into virtuous and *ignea* water, *Aqua vitae*, which contained the essence of the medicines. The Idea of quintessence is also present in Arnald of Villanova's contemporary work, Ramos Lull (1232-1316). See Alfonso-Goldfarb (1987, pp. 148-151).

[6] About distilling, see Roxo Beltran (2000, pp. 23-27, 32-43).

Many British authors found out about chemical philosophy, also known as paracelsism or Paracelsus' philosophy,[7] from Thomas Erasto, who published four parts of his *Disputationes de Medicina Nova Paracelsi*[8] between 1572 and 1573.

Paracelsus advocated a theory based on a concept of radical illnesses different from the humoralists, which caused medical science to expand amazingly. Paracelsus insisted that the reason for the sickness was not an imbalance of the humors of the body, but a specific case, outside the body (Borstin 1989, pp. 312-314).

According to Paracelsus when God organized the universe, he created a remedy for each disturbance (Paracelsus 1999, p. 117). The causes of the sicknesses were closely related to the minerals and poisons brought from the stars' atmosphere.

In the 16th Century, there paracelsists were convinced that they were living in a new violent age, a time in which unknown and devastating sicknesses came from the "New World". There was a demand for more powerful medicines than those developed by the galenists – the only ones the medical schools accepted at that time – made them reject the knowledge being spread by the medical schools, considering them out dated and deadly.

However, it was not only the two forms of Physic that were an issue in the 16th and 17th Centuries,[9] according to many science historians' opinions, the "scientific revolution" which started at that time argued about the three different ways of seeing the world, the Aristotelian one, the magic one (which included alchemy and iatrochemistry), and the mechanicist one (Hudson 1994, p. 35).

The substances were considered bearing two types of property: the elementary ones, which are responsible for their more perceptive characteristics such as taste, smell, density, color, etc., and the hidden ones, which gave them inexplicable properties, those that were only controlled by the creator, such as medicinal ones or magnetic ones.

---

[7] Because was Paracelsus, who compiles that teory, and was its most zealous defender.

[8] Here, Erasto considered Paracelsus a dangerous innovator, who prescribed lethal potions as medication. Paracelsus' use of magic was the reason for Erasto's comparing him to the devil.

[9] Humoralistic medicine and chemical medicine.

The hermetic writings only became known in Europe during the fall of Constantinople in 1453, and were the basis of the magical tradition of European Science. This influence of the Alexandrine alchemy added a more mystical character to the work of the neo-platonists.

Many of the alchemical works of that period were iatrochemistry. Paracelsus had used the neo-platonic doctrine of microcosm and macrocosm to medicine. Up to mid-16th Century, many progressive Physicians, despite being galenists wrote iatrochemistry's texts in favor of the use of the medicines which had been chemically prepared. Joseph Duchesne – the *Quercetanus* – (1544-1609) is a typical example.

The paracelsists looked upon the world and nature as a large alchemical laboratory. The terrestrial crust formation could, apparently, be reproduced in chemical instruments.

According to the paracelsists illnesses were due to external causes and were situated in specific organs. They were localizable entities that could be chemically defined and treated by internal *archei*,[10] whose roles were to be like alchemists situated in different organs of human bodies, therefore Physicians should imitate the *archei* so as to heal these illness seeds which were introduced into the body.

When following this philosophy, one understands the value of purifying to a chemical medicine,[11] which frequently took on a religious character. Upon purifying the medicines, the Chemical Physician would also be purifying himself, because "without the *labor pio and saint*, the fasting and meditations the alchemist would never reach purification or the beings' souls" (Bostocke 1585, sig. C.f. (ar), D.f. (r)).

Anima (which is mid-way between the *corpus* and the *spiritum*) would bring unity and movement to bodies. The physician should aim

---

[10] *Archeus*, according to Paracelsus was a type of *vetor*, used astral forces on the inferior plane, a purifier, an alchemist, which situated itself in each part of the human body, possibly, the most important one being the one which is situated in the stomach, separating the pure and the impure, food excretion. For further information, see Alfonso-Goldfarb (1987, pp. 160-163), Pagel (1982, pp. 105-112), Reis (2000, pp. 31-42).

[11] "As the gold that has not been through the fire is useless, also useless and bad is the medicine that has not been purified by the fire. Because everything has to go through the fire to achieve a new birth, which is useful to men", Parcelsus (1999, p. 130).

at strengthening the *anima*, so it could react to the bodies' hurts and passions, it would only be nourished by fluid which has become ethereal by fire, once strengthened "it could carry out its tasks and deeds, in peace and unity, as God wishes it to be and not with the disharmony and controversy of pagans" (Bostocke 1585, sig. C.iii. (av), C.iii. (ar)).

Although there have been uncountable contradictions to Paracelsism, in English, since the 16th Century, the opposite of what occurred on the continent, where the followers of this idea did not cause disputes with the Galenists and often did not even take a stand, neither in favor nor against Paracelsus.

On the continent, and later in England, iatrochemistry was able to gather keen followers and obstinate enemies. These passions set off a debate which culminated in helping to advertise chemical medicine as early as 1618. When the English Pharmacopoeia was published, various sections printed about how chemical medicines were prepared.

If the Galenists' way of healing was based on the principles of opposites[12] to sickness, the chemically prepared medicines stuck strictly to *similis*[13] theory, where worrying about purification and dosage were constantly present in the works of Paracelsists.

Chemical physicians hold the opinion that man was created of the same *materia* (matter) with which God created the whole Universe,[14] the *Illiaster*, and the human body – the *microcosmus* – were also structured according to the Universe – the *macrocosmus* – and to take

---

[12] If the patient were to show symptoms, which lead the physician to believe that he was suffering from an excess of any of the humors: heat, cold, dry or humid, he would be treated with the opposite humor.

[13] The popular Germanic tradition – like many others – suggested that the cure should be found by means of similar principles. Paracelsus adhered to this theory, which thus became one more characteristic trait among his followers.

[14] "Like the sun and the moon are separated between themselves, even though, in ancient times they were only one, so health and disease were only one thing, which later were considered separately, like the moon and the sun. And as these rise and in the huge sphere of the sky, one appears then another, as do the stars – and one should know - they are entwined with man's body and similarly divided, the same happens with all health and disease manifestations. Because all must be present in the body, so that the 'internal firmament' can be complete and carries out the number of its parts." Paracelsus (1995, p. 121).

good care of man's health one must understand the Universe (Paracelsus 1995, pp. 126-133).

Diseases like dropsy, leprosy, and syphilis seemed to be under better control with the chemically prepared medicines and this was even stated, by Paracelsus' most fervent opponents; however, his philosophy was neither accepted nor publicized in British territory. For a long time, Paracelsus was considered one of the typical representatives of the "magic" tradition and, was sometimes placed in radical opposition of the "true men of science". But authors like Walter Pagel and Charles Webster, stated that this would be "historically absurd". According to Webster, the "Natural Magic" made laboratory work more valuable from many angles and carried an "animist" philosophy found in a very similar way in works written by Kepler, Gilbert and Harvey (Webster 1982, pp. 20-29).

It is not easy to state, what the points are, and that could be considered new or from the ones "discovered" by Paracelsus. It is equally as difficult for Paracelsus' reader to avoid the idea of connecting many of his medical propositions to the works carried out by modern science. At a conference held in Paracelsus' home town at a ceremony for the four hundredth year of the latter's death, Carl Gustav Jung, refers to this fact in the following way:

> Perhaps I should ask my readers to forgive me for the heretical idea that Paracelsus would presently and undoubtedly be the advocate of all such arts that the medicine presented by the university prevents from being taken seriously, that is, osteopathy, magnetopathy, ocular diagnosis [iridology], several food mono-manias [anthroposophic medicine], sorceries, etc.[15]

He also adds:

> I would also like to point out an extremely important aspect of his therapy, the psycho-therapeutical. Paracelsus also knew the very ancient method of *conversation about the disease*, of which *Papiro Ebers* had already provided us with so accurate examples from the time of Ancient Egypt.[16]

---

[15] Jung, "Paracelsus as a Physician", in Paracelsus (1999, p. 263).
[16] 16th Century. See Ebers, *Papyros Ebers. Das hermetische Buch über die Arzneimittel der Ägypter. Apud.* Paracelsus (1999, p. 277).

Therefore, Paracelsus' concept on diseases involves both the body and the soul, in such a way that Jung, in his book "Psychology and Alchemy", seeks in alchemy and in its transmutation sequences,[17] the "process for individuation of a human being".

Walter Pagel does not discard such comparisons, either, which is made clear in his book *Paracelsus*, where he says that Paracelsus has provided Medicine with sharp observations about diseases and pathological conditions, such as the works about the miners' disease, the first work attempt in "Occupational Medicine". He studied potable and mineral water, the treatment for dropsy and cretinism; he also recommended the use of mercury as diuretics and showed the presence of albumin in urine. He also tried to insistently implant a new pathology system, such as that of tartar,[18] and attributed great importance to the external pathological agent.

This pathology system points towards the direction of a modern trend where diseases can be distinguished as "objects" that can be classified and isolated due to anatomical transformations that are peculiar to them, thus allowing their being treated as specific causes.

We had better check the words of Pagel himself:

> In short, Paracelsus not only demolished the ruling system of medicine, but replaced it by a theory in which the germ cells of modern pathology can by divined. How far he was ahead of his time [...]. When he pointed towards the outside of the body, when he insisted in the uniformity of the causes and in the specificity of diseases, he was showing the way to modern medicine. If his arguments were in-

---

[17] Refers to the alchemic transmutation processes: *Nigredo, Albedo, Citrinas* and *Rubedo*. Jung. "Psychology and Alchemy", pp. 240-244.

[18] According to Paracelsus, tartar is the excrement of beverages and food. Caused by "Man's spirit", it is only produced by the "salt spirit" – the only one of the three principles to have the *lapidia materia*, the stone matter- over which human heat acts similarly to the heat of the sun, drying all the mucilage and viscosity, coagulating it (solidifying it). This salt coagulation, the tartar, can place itself in several parties of the body, such as the stomach, the intestines, the diaphragm, the liver, the kidneys or in the mouth. [It is interesting to refer here to the description made by Paracelsus about the damages that tartar could cause when placing itself in the mouth, for example, adhering to the teeth]. It would cause "gums putrefaction", holes, toothache and other similar ones, due to tartar's acrimony (acid nature)". Pagel (1982, pp. 153-165).

accurate, his criterions and his intuition were surely correct (Pagel 1982, pp. 345-347).

We will certainly find theories and quotations in Paracelsus's works that will refer us to contemporary medicine. It is a dangerous position that calls for redoubled attention on focus and contextualization.

One of the first books that accepts and spreads Paracelsus's theory in England is *The Difference...* Having been published in London in 1585, this work is considered by some science historians as an actual apology to Paracelsus' work and, possibly is the first work in English to disclose such philosophy (Debus 1996, p. 57).

In his book, Richard Bostocke, na Esquire, who belonged to John Dee's study group (Harley 2002, p. 1), intends, besides highlighting the qualities of chemical medicine, to show that it is the most traditional, the best and most coherent with God's rules...[19]

Bostocke entered St John's College, Cambridge, as a Pensioner,[20] in the Easter of 1554, the same college John Dee had joined in 1542 and enrolled two years later, in 1554 (Harley 2002, p. 2).

J. Venn and J.A. Venn, in *Alumni Cantabrigienses* (Venn & Venn 1872, p. 12), suggested that Bostocke might have been the same Richard Bostocke from Southwark, who had been admitted to the Inner Temple in February 1551. Nevertheless, according to the historian David Harley, the link to Southwark still remains obscure (Harley 2002, p. 2).

A descendent of an ancient family of Cheshire's lower aristocracy, Bostocke was appointed to Tandridge's court, in Surrey, in 1554.

He served as a Bletchingley Parliament member for four times, between 1571 and 1585, where he possibly entered through the influence of Baron Howard of Effingham, a friend of one of R. Bostocke's clients, who, presumably put in a word for him. His mother, Foelice Heaton, had also been born in Bletchingley (Harley 2002, p. 3).

---

[19] Webster (1982, p. 21), includes a comment by Newton, who shows how much he also marveled at this fact "the admirable new paradox that alchemy should be in concord with Antiquity and Theology", I. Newton, *Manna*, Microfilm of the University of Cambridge's Library. Keynes MS 33, f. 5r.

[20] At University of Cambridge: as a student that pays for his studies working for the University itself.

In order to claim that Rychard Bostocke had a close relationship with John Dee, more research would be necessary; nevertheless, in the loan record catalog Dee kept, the use of a copy of *Oviedo* is recorded, a history about West India,[21] written by R. Bostocke.

The author of R. Bostocke's biography, historian David Harley, ponders that it might have been Mr. Holton, physician and parish priest of Oxted, Tandridge, and English orthodox Calvinism adept, one of R. Bostocke's clients, who influenced the explicit religious enthusiasm inherent in all his works (Harley 2002, p. 3).

John Dee's ledger records the fact of his having been in Tandridge and dined with Mr. Oxted, in 1582 (Harley 2002, p. 3), which leads us into thinking that there had been a meeting between Dee and R. Bostocke, when they could have discussed about his *The Difference*... In September of the following year, Dee leaves France, taking along his alchemic and Paracelsus' books.

In chapter seven of *The Difference* "One cause why the Author did write this treatise", R. Bostocke states:

> I was the last Parliament time before this that is now sommoned at the table of a reverend Bishoppe of this land, which was not unskilfull in Phisicke, in the company of a Phisition, which inveying against this aucient Phisicke, by the name of Paracelsus his Phisicke, ignorantly attributing to him [Paracelsus] the first invention thereof, pleased himself and someof his audience, in telling that the same Phisicke, had no ground nor foundation, neither any being (Bostocke 1585, sig. D.f.(r)–D.f.(v)).

Through Bostocke's comment, one can presume that the inspiration for the book occurred during the third Parliament session, which lasted from 1572 until January to March 1580/81. His book was written before the parliament convened again on November 23, 1584, in London, when Bostocke would have presumably brought his manuscript and had it registered at the Stationers' Company, on December 7, 1584 (Harley 2002, p. 4).

---

[21] John Dee kept two copies of all his books about magic, cosmology and chemical philosophy, and one of such copies was kept aiming at the dissemination of these philosophies, insofar as they were lent to his contemporaries; such loans are recorded in his "Loan Catalog" in Cambridge. See Harley (2002, p. 3).

So as to understand why that work has been so little disseminated, many investigations shall be carried out yet, amongst which, the number of published issues and how their distribution was effected.

The manuscript is at the University of Cambridge, signed by R. B. Esquire Thomas Lorkin, the illustrious Medicine Professor of University of Cambridge, whose stately library used to serve the University itself, who owned, in his private collection, a volume of R. Bostocke's book, among other works by Paracelsus, and by the English and European supporters of Paracelsus' work.

His dense work, especially when he deals with medical chemistry is full of quotations from Paracelsus' works, even though they are not pointed out as such, but rather appear blended with the author's opinions. All the work was developed having History as its background, with the intention of granting to iatrochemistry a cabalistic authority.

Bostocke's marked religiousness, his being a puritan, makes him place Adam's fall as the first cause of diseases, because with his "sin" he brought about impurity and corruption to all the things of the world, "not onely in man, but also in all living creatures, Hearbes, Plants, Mynerals: and in the fruites of the firmament and ayer"(Bostocke 1585, sig. B.iiif(v)). If, it was due to the breach of the obedience of man towards God that the unity was broken and impurity was united to purity, R. Bostocke could only praise a medical philosophy that brought about in its pillars, the concern about purification through fire or the quintessence extraction, the accurate dosage and that would treat diseases by means of the *simili* theory.

Like Paracelsus, R. Bostocke believed that Man had been withdrawn from the first *Matrix*, the earth's core, from which the great world had its origin and, together with other creatures, it was modeled by God's hands. Man had, therefore, an "earthly" body, or *animal*, which was constituted of exactly the same material that the whole Universe had been constituted. Such a body would be formed by the "elements" Earth and Water, and "inhabited" by a "sidereal" body or "spiritual", formed by Fire and Air.

This paracelsists opinion would justify R. Bostocke's acquiescence to the *simili* theory, since he talked about the unity God had provided the world with. In the quotation below, R. Bostocke, when talking about the three distinct medicine types, also puts forward his indignation towards the treatment using opposing principles to that of the disease. To wit, in his own words:

> This auncient and true phisicke consisteth of Medicines of two sorts. The first is *universalis* or *unarii*. The second is *ternarii* or *particularis*.[22] These two are founded upon the Center of unitie, concord and agreement, their scope and is to bring the sicke person to unitie in himselfe, they doe agree with the rule of Gods worde, they depend upon the fountaine of trueth. The Ethnikes or heathen have of their own braynes devised a third kinde of phisicke or Medicine which is *binarii* or *vulgaris*. This is most grosic and worst, and is that phisic is which is most commonly used, and most stoudly maintained and defended. This Phisicke is founded upon a contrary Center to the other therefore a false Center. For it consisteth in dualitie, discord and contrarietie. It maketh warte and not peace in mans bodie (Bostocke 1585, sig. B.f.(v)).

Once more according to Paracelsus, Bostocke also believed that, through the anatomy of the plant and the anatomy of the sick part of man's body, the devoted experienced physician could choose the correct medicine. "Thus, by means of the correct concord of those two anatomies, of the disease and of the medicine, the real cure was developed and accomplished" (Bostocke 1585, sig. B.f.(v), Paracelsus 1999, pp. 154-155).

In regards to iatrochemistry, undoubtedly chapter VIII: "Certaine difference, betweene the Aunticent Phisicke and the Phisicke of the Heathens", from R. Bostocke's work, has special importance. Comprised of nineteen items, in which R. Bostocke reports the ideas and the behavior of the pagan philosopher and, then, makes a comparison with the chemical philosopher. This chapter brings forth the synthesis and the comparison of those two medicines.

Both the theological and the philosophical issues, which are the discordant basis for those two trends, and the issues related to Chem-

---

[22] The main difference between *ternarii* medicine and *universalis* medicine lay in the fact the latter was performed by a very few men, who had been granted the "gift" of medicine by God, whereas the former – *ternarii* – was that medicine performed by ordinary men who studied, devoted themselves to research, worked diligently, seeking to analyze nature. Such physicians also worked very well, since medicine stemmed from all such search and devotion, giving the physician the ability to carry out his work with the same result of the *universalis* medicine. Bostocke (1585, sig. C.iii.(r)).

ical Phisition, are there treated one by one; the need for purification, the *vulcanus*, the *archeu*, the *tria prima*,[23] the *simili* principle, the submission to God and to nature, the repudiation to the scholastic tradition, the *ocultum* and the *manifestum*, the macro and the microcosm, the "seeds of diseases", the tartar disease, the need for acquiring knowledge with the work through fire, ultimately all the main themes of Paracelsus' philosophy are condensed by R. Bostocke in extremely didactic format in this chapter (Bostocke 1585, sig. D.ii.(v)–F.i.(r)). Summing up:

As the first and utmost difference, R. Bostocke places the philosophical basis; the humoralist physicians were centered within a pagan philosophy that descended from Aristotle, whereas chemical philosophy, supported by the chemical physicians, was based on the Gospel and on nature.

Ethinickes Philosophers placed nature as the primordial efficient cause of all the things and God as the secondary cause, whereas chemical philosophers supported that the nature of things was just an instrumental cause, which was totally dependent upon Good's will.

Ethinickes Physicians used medicine in its gross form, without purification, since they did not know this art; the Chemical Physician, however, taught this art and withdrew from such medicines the "ethereal and celestial virtues", before giving them to the sick

Ethinickes Physicians based themselves on the false center, the *binarii* and the *dualitii*, that only gave rise to discord and dissension in human body and they taught that diseases should be cured by their opposites, whereas the chemical physicians based themselves in the real unity center, the *unarii*, which only gave rise to harmony and concord. When this harmony was broken and the correct proportion

---

[23] It is regrettable that Paracelsus was not very clear when dealing with the three principles and their interconnections. According to A. Debus, the characteristic individualism of many of his followers does not make such understanding any easier. Nevertheless, there is concord among historians that sulphur was the cause for combustion, structure and substance, whereas solidity and colour were achieved through the salt and the steams were achieved through mercury. The components, the fuel, the steam and the solid were frequently shown through the action of burning a branch, when the smoke represented the mercury, the flame the sulphur and the ashes the salt. Paracelsus, *Die 9 Bücher de Natura Rerum. Apud.* Debus (1996, p. 28).

between the three principles, the salt, the sulphur and the mercury was annulled, such harmony would be restored using the very "substance" that was in disharmony, not its opposite. Medicine should bring the diseased body to the domineering unity, equally, in the three "substances".

Ethinickes Physicians' followers based themselves just on Gallen's, Avicena's and others' doctrines. Chemical physicians based themselves on the Word of God and on laboratory work, where they learnt about the three substances that comprised all the things of the Universe, about the macro and microcosm, besides learning to purify and to express the *simple* – the *arcana*, the quintessence – through the action of the fire. All in all, they learnt their work on their own, not through books and descriptions of third parties.

Ethinickes medicines followers dealt with the "Simple; as Hearbes, Plantes, Rootes, &c. father themselves upon Gallen, Mesue, Dioscorides, &c" (Bostocke 1585, sig. D.f.(av)). Chemical physicians purified through the fire to extract the virtues of the *simple*, since they knew, by experience, that "in Honey a venomous tartishnesse, and much filthinesse in Sugar. And in Arsenicke excellent good medicine [...]. As the Rosine of one Countrey is not of that nature as the Rosine of an other Countrey" (Bostocke 1585, sig. D.f.(av)).

According to Bostocke, humanity has always known of the danger presented by certain foods and soils that had such a high degree of impurity, that they became highly damaging to man. Due to such knowledge, "our forefathers did diligently viewe and search before they layde the foundation of their houses, townes or Cities, and that they did oftentimes, because they would know whether those parts were infected with disease, with their food and by the often sight of them, they judged whether the Cattell fed there where holsome for their rictual and food or no" (Bostocke 1585, sig. B.f.(av)).

Bostocke mentions as an example a local, "neere the River *Potheus* in the Ile of Candie, neere the Cittie *Cortina*, wherein groweth such grasse and hearbes, that the Cattell which be fedde therwith will have no apparent Spleene" (Bostocke 1585, sig. D.f.(ar)).

Man, when feeding himself would be absorbing poisons and even death itself; however, God has granted each creature with an inner alchemist – the *archeus* – able to separate the pure from the impure in the food they would ingest, but if the individual were ill, the inner alchemist work would not be enough for such a separation, thus, the physician should prepare and administer a free-poison medicine, making its absorption easier (Bostocke 1585, sig. B.f.(av), Paracelsus 1999, p. 171). The chemical physician would be the one who would have the ability to separate the pure form the impure, through the action of the fire.

Bostocke, similarly to the way Paracelsus thought and acted, believed that the physician should travel, should seek to understand the diseases and the medicines used in other countries and regions, since in the "*Labyrinth* and in the *Ars Signata*, by Paracelsus, the latter made clear the variation of the proprieties of things" (Bostocke 1585, sig. D.ii.(ar)). Bostocke quotes numberless examples of medicines still being used nowadays by popular medicine in the form of teas, macerations, unguents and others.[24]

In the sequence of chapter eight of R. Bostocke's work, we will be working with comparison number seven, that he makes for us. Even though the author is rather repetitive in his analysis, he always adds new datum. In this item, Bostocke states that pagan physicians themselves, when writing about the *simple*, pointed out that the importance of the nature of things, expressed by "causalities", which perish in the digestion and, because of this, they disseminated the use of Herbs, which "blind the physicians". Nevertheless, chemical medicine judged nature's things only after annulling all its external features, just leaving their *arcana*.

At this point, it is interesting to make use of R. Bostocke's quotation to analyze some aforementioned drugs, which were probably of current use:

> For mortification is the beginning of dissolution and separation of good from evil [...] Aron, called in english *Ceccowpint,* hath a very hotte taste in the leaves and roote.Wormewod hath a bitter taste, yet by light digestion, preparation and separation of their vertues and propertties from their bodies, they vanish away and be lost. It is otherwise in Ginger, because his heate is stable, lively and solided in his naturall seede, vertue or propertie, and cleaveth to it stedfast. The *vitrum* of

---

[24] As an example, chamomile, in the form of tea, as a functional stimulant to the stomach; mint, macerated with oil and used as scar-healing, and others.

> Antimonie is wthout any taste, yet for all that is vehement Purgation. Lead likewise hath no taste, yet not withstanding,[25] a pleasaunt sweete Sugar wilbe drawen out oft, comfortable or pourging medicines, or such as cause sleepe, can never be found out by their taste of heate or cold. [...] that is *occultum* and *manifestum*, and that *manifestum* is commonly contrarie to *occultum*. By this means of the fire, they find Quicksilver in manifesto is cold and moyste, and within his occultum, is hotte and drie [...] the *Arcaum* liveth on the *occultum*. (Bostocke 1585, sig. D.ii.(av)–D.iii.(ar))

Always comparing "Art" and nature's work, Bostocke says that, equally to the seed that has to suffer putrefaction to be reborn for a second life when launched into the soil, and it is this second life that is profitable for the plant, the stomach digests, through putrefaction all the things deposited in it before making use of them, so shall the physician behave: extract the *arcana* from the medicines for them to be useful for the patient.

The next comparison between the two medicines, which is offered to us by R. Bostocke, deals with the humors. He sees, as the main difference in this field, that pagan medicine attributed Man's build to few humors, whereas chemical medicine attributed his own humors to each member, in accordance with his physique and with each humor's effects; he did not believe in various degrees of heat, cold, humidity or dryness, since they could not be proven, as the three principles could.

In item nine, the chemical philosophy's comprehensiveness and Paracelsus' "cosmogony" acceptance can be realized when R. Bostocke insists on the need to understand the macrocosm. According to him, pagan medicine used the humors and the qualities to identify Man's diseases, whereas chemical medicine knew that, if the physician wanted to know Man's nature, with all his diseases, he would first have to know the diseases that all the things in nature undergo in the "great world", by reducing such bodies to the three substances. In a determined macrocosmos place, the physician would identify one disease or the other in a dispersed way, whereas, concerning Man, all of them could be seen. Nevertheless, this knowledge could only be acquired if the physician knew how to resolve chemical philosophy,

---

[25] Lead, in the 16th century was considered an impure metal, decayable, that could not resist well to time action; such metal "derivated" from impure gold.

"for *Chymia* and *Medicina* may not be separated asunder, no more than can preparation or separation from knowledge or science" (Bostocke 1585, sig. C.f.(r)).

Ethnickes medicine and its followers attributed diseases' causes to dead qualities: cold, heat, humidity and dryness, "So they make no difference between fire and smoke, between seeds and their fruits, between substances and their accidents, between the thing itself and his excrements" (Bostocke 1585, sig. C.f.(r)–C.f.(v)). However, the chemical physician:

> [...] proveth, that there bee spirituall Seedes of all maner deseases, indowed with lively power, which bring forth those qualities, and all other fruites of deseases, and their sundry kinds of griefes in our bodies, as and earth bringeth sorth fruite by meanes of seedes in it. And that those qualities be onely figures, colours and *Symptomata* of deseases. And though each desease be either hot or colde, &c. yet they be but signes and conditions of desease, and not the desease it selfe. But all deseases are in the three substanties of *Sal, Sulphur*, and *Mercury* (Bostocke 1585, sig. C.f.(r)–C.f.(v)).

It is, perhaps, in item eleven that we shall find the larger number of "action forms" of the medicines used at that time, actions which Bostocke called "mechanical spirits"; he stated that Ethnickes physicians prepared their medicines according to the "accidental qualities", already mentioned; however, they themselves used hot things before the cold things, without worrying with their qualities, for example. "But the followers of the Ethnickes themselves in taking one hot thing before one other, as Pepper before Chamomile, &c. and one cold thing before one other, does testify sufficiently that they sought not heat nor cold, but *Arcana* which they thought were degrees" (Bostocke 1585, sig. E.ii.(r)). The chemical physician, however, Bostocke says, sought the medicines' virtues, their essence and took into consideration the nature of things, rather than their humors or qualities, but indeed their seeds and their mechanical spirits, that is, whether they were:

> *Attrativa, Anodymæ, Abstergentia, Aperitiva Constringentia contrahentes. Quale membrum principale respicientia:* or in *Carne cartilagine ossibus sanguine Synovia, &c. operantia condensantia, conglutinantia, Corrosiva confortantia, coagulantia digestiva diuretica, dyaphoretica, dormire facientia discussiva expulsiva eracuantia, extenuantia, famen moventia gravedinem moventia, horrorem*

moventia, renovantia incidentia, incrassantia, inflamantia, incarnativa mundificativa mollificativa maturantia, mortificativa, morbus quosdam respicientia Martialia Narcotica Nitrosulphuræ nutritiva oppilantia purgativa penetrativa, retentiva regenerantia repellantia repercussiva resolventia trahentia ulcerativa venenum repugnantia vomitum morentia and such like (Bostocke 1585, sig. E.ii.(r)– E.ii.(v)).

In this item sequencing, Bostocke talks about the use, by the humoralist of the Colocynth as purgative and says that Dropsy expels the *Sal Resolutum*, without caring or worrying about the qualities. "So in all purgations the Ethnicks themselves are driven away from their accidental qualities [...], that is why the followers of the Heathen often tymes, yea after their consultations, either know not what to do, or else determine often the worst rather than the good for their patient" (Bostocke 1585, sig. E.ii.(v)).

Human bodies' disease is compared to the metals' diseases, both regarding the gold "deterioration" inside the mine and concerning the circulatory problem that can lead to gangrene; the proposed solution is to seek the balance of the three principles, since the excess of one of them is what would be causing the blockage of the human balsam circulation. Mummy[26] or balsam would be able to renew and allow the healing, rather than the cutting or the traumatic treatments.[27] Due to the same reason, bleeding is unadvised in most cases, since it would make the patient lose the natural balsam; in this case, what should be used for blood renewal would be the *"potiones vulnerariæ and consolidantia"* (Bostocke 1585, sig. C.iij.(r)).

In item twelve, R. Bostocke again compared these two medicines, stating that Ethnickes doctrine kept itself contemplative and fought against experience, whereas chemical doctrine based itself in the experience associated to the knowledge of each thing's properties, virtues and nature. Thus, the latter would know "what man is, what the med-

---

[26] According to W. Pagel, the Mummy or the Balsam, could have been the Salt, rather than the chemical substance, but in the sense of the natural power of the tissue cure, preventing its putrefaction. Pagel (1982, p. 101).

[27] The chemical physician was against the surgery using cuts. R. Bostocke uses the term surgery to designate the balsamic process and treatment with medicines associated with food diets and specific exercises; it is clear that he did not approve of the surgery performed by the "barber surgeons" who he considered aggressive and which he thought caused disharmony in Man's body.

icine is, how they agree in right *Anatomia* [...]. So that experience and practice ought not to proceed of speculation, but speculation ought to be derived out of practice" (Bostocke 1585, sig. C.iij.(v), Paracelsus 1999, pp. 99-104, 157-158, 171).

In this explanation R. Bostocke highlights that humoralist feared that the medicine prepared by fire kept all the caustic corrosive qualities of this "element". However, Bostocke recalls that fire consumes just the corruptions so as to restore the natural unit that had been consumed by the *elementary fire* (Bostocke 1585, sig. C.iiii.(r), Paracelsus 1999, pp. 171-174). He also mentions the "unjustified" fear of using metals as medicines, since the chemical physician worked the metal before using it as a medicine and, therefore, they would be no longer metals or minerals in their essence, but indeed "volatile spirits" (Bostocke 1585, sig. C.iiii.(v), Paracelsus 1999, pp. 157-158).

Chemical physicians, according to the analysis made by Bostocke in item thirteen and fourteen, contrarily to the Galenists, named the diseases according to their features and not just because of the imbalance of their humors. Thus, knowing all the diseases' characteristics, iatrochemical physicians were more capable of choosing the medicines that should be used, which were more coherent with the diseases' anatomy. He mentions some examples of diseases' names that he considered coherent, such as the sulphur disease, the Peter Kendled salt, the several diseases that are called tartar diseases, due to the similitude that such diseases had with the tartar salt, that is, having given the correct name given to the disease, it would be better understood (Bostocke 1585, sig. C.iiii.(r)–C.f.(av)).

After pointing out once more, in item fifteen, the importance of the chemical physicians' ability in ministering the purified medicine in small doses, correctly balanced for each patient, R. Bostocke states, in item sixteen, that chemical physicians observed the signs and tokens of diseases and classified them in accordance with their origin. They would comprise two types: *Cælestes* and *Terrestres*. Thus, those diseases, the origin of which was in the superior world could not be removed with vegetable *arcana*, since these "required high degrees of preparation", and he goes on:

> Likewise if cause of disease proceede of Mierals or metals, they must be cured with *Arcana* of Mynerals, because such will not feeld *Arcana*

of vegetables, that is of hearbes and rootes, &c. But if the disease bee caused by influencies of heavens, neather of the other *Arcana* will serve, but they are to be cured by Astronomy and influencies. But those Diseases and griefes that come by supernaturall meanes, will not be holpen by any meanes aforesaybe, but by supernaturall meanes (Bostocke 1585, sig. E.ii.(ar)–E.ii.(av)).

Bostocke also praises the chemical physician's concern regarding the existence of diseases which are more impetuous, more painful and more persistent than others and that such differences and the capacity of removing those diseases would consist in the knowledge of their origin, their roots and, consequently, in the use of the most relevant *Arcana*.

Like Paracelsus, he criticizes, in item seventeen, the judgment of diseases just through urine observation, and says that such judgment could only be made after the due separation of the urine by fire, since:

> So shalbe see the matter of each disease and his medicine and touch it with his hande, whereby his shalbe able to give a perfect judgement if he bee able to judge as becommeth a Phylosopher and a Physytion. By this means shall bee, finde the urine not bee *Meretrix* nor lyar (Bostocke 1585, sig. C.iij.(v), Paracelsus 1999, pp. 107-108).

In the last item to be dealt with in this chapter, R. Bostocke shows that he believes that "Ethnickes Phisition" had not been able to adapt itself to new diseases and thus allowing it to be updated.

> The followers of Ethnickes in their medicines, credite *Recipes* of Gallen, Avicen, and such other, though in these daies the bodies of men bee not so strong as they were in their time. And though diseases in nature doe daily alter (Bostocke 1585, sig. C.iiii.(ar), C.iiii.(av), Paracelsus 1999, pp. 113-114, 129-133).

Bostocke thought that Ethnickes Physicians were unfair because they criticized chemical medicine without knowing it. And he says that they had better study it and understand that art, before criticizing it. In this part, the huge importance R. Bostocke lent to laboratory work is shown and also shows that he, albeit a layman in Medicine and Alchemy, knew[28] the extraction and purification used in this art as well. To wit:

---

[28] Although to state that R. Bostocke had dealt with such methods, deeper researches would be necessary, he, anyhow mentions them with conviction and pertinence.

> Which of them knoweth what way to begin to separate the Salt, Sulphur, and Mercury, from Hearbes, Plants, and all other thinges as it ought to bee artificially, according and agreeable to properties and several Hearbes require severall maner of separations. Plantes have their peculiar separations: Mynerals theirs: Marchesits theirs, &c. Which of them doth knowe the severall maners of *Calcination, Reverberation, Cementation, Incineration, Imbybytion, Pastation, Liquefaction, Ablution, Sublymation, Exaltation, Contrition, Resolution, Putrifaction, Circulation, Inhumation, Distillation, Ascention, Fixation, Lavation, Coagulation, Asfation, Congelation, Fermentation,* &c. And the natures and properties of these severall works and operations, wereby *Regeneration, Tyncturs, Arcana, Magisteria, Quintum esse,* and *Elixirs* be had and gotten. Wich of them can tell what transmutation of Elements meanth? (Bostocke 1585, sig. F.i.(v), F.ii.(r))

All this indignation leads Bostocke into criticizing once again those who made long speeches against chemical medicine just by having read or heard criticisms about it. They, who "merely follow their prince and captain Gallen" (Bostocke 1585, sig. F.ii.(r)). He regretted that those "ignorant people" were defended and protected by the authority of princes, even when they "destroyed" their patients.

> [...] and they are excused and discharged of their fact, by the lawe called *Lex Aquilia* [...] in the Scholes to be heresie and soule ignoraunce to speake against any part of Aristotle, Gallen, Avicen, or other like heathens doctrine [...] Wherefore if the Chymicall doctrine agreeing with Gods Worde, experience and nature may come into the Scholes and Cities in steade of Aristotle, Gallen, and the other heathen and their followers (Bostocke 1585, sig. F.iii.(v), F.iiii.(r)).

Prior to R. Bostocke's book, there existed some translated works by John Hester and G. Baker. John Hester, like Baker, was also a close friend to Thomas Hill, who bequeathed him his second book, which was a translation into English by Joyfull Jewell, by the Italian physician Leonardo Fioravanti (Debus 1996, p. 65).

Hester was a practical businessman, highly knowledgeable in "minerals, herbs and flowers", besides being an active distiller in London, around 1570. After Joyfull Jewell, in 1579, Hester does not stop translating. He translates all the works by Fioravanti, Duchesne, Hermann and several spurious works by Paracelsus. Having no interest in the deeper aspects of Paracelsus' theory, he preferred works which contained little theory and large amount of prescriptions.

Although he had made translations of undeniable importance for the introduction of chemically prepared medicines in England, his work did not have the same importance concerning the dissemination of Paracelsus' ideas. Nevertheless, according to studies done by A. Debus, no other pharmacist or distiller was as straightforward in his praises to chemistry as Hester was.

For the first forty years of the 17th century, besides the book by Bostocke, those were the sole existing works in English about Paracelsus theory, until J.B. Helmont's appearance, which brought forth a new interest to Paracelsus' writings.

As of the W. Pagel and A. Debus researches, the map of Paracelsus' theory has been combed in detail. Meetings, papers and publications have been showing a thin capillary thread, which joins Paracelsus' iatrochemistry to ancient and medieval conceptions and that extends itself up to the limits of modern medicine in the 19th century.[29]

Nevertheless, through studies like the one by Debus, we presently know that this compound iatrochemistry cannot be sectored, at a single time, through medical and theological methods. Its power, by the way, would lie exactly on this composition that, albeit of difficult analysis for a contemporary professional, has provided it with a privileged status in the development of modern medicine.

The Difference Betwene the Auncient Phisicke…and the Latter Phisicke, the dense text written by R. Bostocke in the 16th century, places this author within one of the most important knots of the fine historical thread that helped established Paracelsus' iatrochemistry at places where modern science would gain force: England.

References

Alfonso-Goldfarb, A.M. (1987), *Da Alquimia à Química*, São Paulo: Nova Stella editorial/Editora da Universidade de São Paulo.

---

[29] Refer, for example, to the collection of works published by Debus & Walton (1998). In October, 1999, in St. Louis, every History of Science session dealt with the subject; in Brazil, for example, besides the works around the subject carried out by CESIMA's study group, there are those specific ones written by Paulo Porto, which are Porto (1998), Ruiz (1999), and the MA dissertation of my authorship (Reis 2000).

Alfonso-Goldfarb, A.M. (1991), "Atanores, Cimitarras, Minaretes: cultura árabe como tecido do saber sob o céu 'medieval'", *Revista da SBHC* 5: 33-40.

Alfonso-Goldfarb, A.M. (1999), *Livro do Tesouro de Alexandre: um estudo de hermética árabe na oficina da história da ciência* (translated by Safa Abou Jubran and A.M. Alfonso-Goldfarb), Petrópolis: Vozes.

Alfonso-Goldfarb, A.M. (1994), "Questões sobre a Hermética: uma reflexão histórica sobre algumas raízes pouco conhecidas da ciência moderna", *Cultura Vozes* 4: 13-20.

Amundsen, D.W. (1968), *Medicine, Society, and Faith in the Ancient and Medeval Worlds*, Baltimore: Johns Hopkins Press.

Beltran, M.H.R. (2000), *Imagens de Magia e de Ciência: entre o simbolismo e os diagramas da razão*, São Paulo: Educ/Fapesp.

Borstin, D. (1989), *Os Descobridores: de como o homem procurou conhecer-se a si mesmo e ao mundo*, Rio de Janeiro: Ed. Civilização brasileira.

Bostocke, R.E. (1585), *The Difference Betwene the Auncient Phisicke, First Taught by the God by Forefathers, Consisting in Unitie Peace and Concord: And the Latter Phisicke Proceding from Idolaters, Ethinickes, and Heathen: as Galen, and Such Others Consisting in Dualite, Discorde, and Contrariate*, Imprinted at London for Robert Walley.

Debus, A.G. (1974), "The Chemical Philosophiers: Chemical Medicine from Paracelsus to van Helmont", *History of Science* 12: 235-259.

Debus, A.G. (1992), "Alchemy and Iatrochemistry: Persistent Traditions in the 17th and 18th Centuries", *Revista Química Nova* 15 (3): 262-268.

Debus, A.G. (1996), *The English Paracelsians*, New York: Franklin Watts.

Debus, A. & M. Walton (eds.) (1998), *Reading the Book of Nature: the other side of the Scientific Revolution*, Kirksville: Thomas Jefferson University Press.

Galen (1994), *On the Natural Faculties*, Chicago/London: Encyclopædia Britannica (Great Books of Western World, Vol. 9).

Harley, D. (2002), "Rychard Bostok de Tandridge, Surrey (c.1530-1605), M. P., Paracelsian propagandist and friend of John Dee",

Indiana Department of History, University of Notre Dame. Paper, available in: <http://www.nd.edu/~dharley/medicine/bostocke-paper.html>.

Hippocrates (1994), *Hippocratic Writings*, Chicago/London: Encyclopædia Britannica (Great Books of Western World, Vol. 9).

Hudson, J. (1994), *The History of Chemistry*, Hong Kong: Macmillan.

Jung, C.D. (1990), *Psicologia e Alquimia*, Petrópolis: Vozes.

Pagel, W. (1962), *Paracelsus: An Introducion to Phisical Medicine in the Era of the Renaissance*, New York: Karger.

Pagel, W. (1968), "Paracelsus: Traditionalism and Medieval Sources", in Stevenson, L. & R. Multhauf (eds.), *Medicine, Science, and Culture*, Baltimore: Johns Hopkins Press, pp. 50-75.

Pagel, W. (1984), *The Smiling Spleen: Paracelsianism in Storm and Strees*, New York: Karger.

Paracelso (1979), *Man and Works, Seletecd Writings*, Princeton: Princeton University Press, pp. 101-140.

Paracelso (1995), *Textos esenciales*, Madrid: Siruela.

Paracelso (1949), *Volumen Medicinae Paramirum*, Baltimore: The Johns Hopkins Press.

Pliny (1989), *Natural History, Libri XX-XXIII*, London: Harvard University Press.

Porto, P.A. (1989), "O Contexto Médico na Montagem das Teorias Sobre a Matéria de J. B. van Helmont", *Tese de Doutorado*, PUC-SP, São Paulo.

Reis, I. de F. (2000), "Recontando a História da Iatroquímica: R. Bostocke e o The Difference Betwene the Aunciente Phisicke... and the Latter Phisicke", *Dissertação de Mestrado*, PUC-SP, São Paulo.

Rattansi, P.M. (1963), "Paracelsus and the Puritan Revolution", *Ambix* 2 (1): 24-32.

Ruiz, R. (1999), "A Montagem da Teoria da Dinamização dos Medicamentos Homeopáticos de Samuel Hahnemammm", *Tese de Doutorado*, PUC-SP, São Paulo.

Venn, J. & J.A. Venn (1877), *Alumni Cantabrigiensis*, Students Admitted to the Inner Temple, 1547-1660.

Villanueva, A. of (1943), *The Earliest Printed Book on Wine*, New York: Schuman's.

Webster, C. (1982), *From Paracelsus to Newton: Magic and the Making of Modern Science*, Cambridge: Cambridge University Press.

Webster, C. (1993), "Paracelsus: Medicine as Popular Protest", in Grell, O.P. & A. Cunningham (eds.), *Medicine and the Reformation*, London: Routledge, 1993, pp. 57-77.

Wellish, H. (1975), "Conrad Gesner: A Bio-Bibliography", *Journal of the Society for the Bibliography of Natural History* 7 (2): 151-247.

Yates, F.A. (1993), *Ensayos reunidos, III. Ideas e ideales del Renacimiento en el Norte de Europa*, México: Fondo de Cultura Económica.

Yates, F.A. (1993), *La filosofía oculta en la Época Isabelina*, México: Fondo de Cultura Económica.

William Bateson's *Materials for the Study of Variation*: An Attack on Darwinism?

Lilian Al-Chueyr Pereira Martins

Department of Biology,
Faculty of Philosophy, Sciences, and Literature of Ribeirão Preto
University of São Paulo (USP)

1. Introduction

The aim of this paper is to discuss to what extent the English naturalist William Bateson (1861-1926) was committed to Darwinism in his early career (till the end of the 19th century). There is no consensus among philosophers and historians of science on this issue. The main focus of our attention will be Bateson's book *Materials for the Study of Variation* (1894), a huge catalogue of facts that substantiated the discontinuity of variation. This work is both regarded variously as an attack on or as a return to the Darwinian tradition by historians and

---

· I am very grateful to FAPESP (Fundação de Amparo à Pesquisa do Estado de São Paulo) that supported this research. I would like to thank Mrs. Rosemary R. D. Harvey, from *John Innes Archives* as well as Mr. Geoffrey Waller from the Manuscript Session of *Cambridge University Library* for their help. I am also grateful to Professor Roberto Martins for his advice and criticisms which contributed to the first version of this article.

philosophers of science. This paper will not deal with the controversy between Mendelians and Biometricians.

2. Opinions on Bateson's position

William Bateson is classified both as Darwinian and anti-Darwinian by different authors including not only historians and philosophers of science, but also his coeval scientists.

The evolutionist Vernon Kellogg (1907) described Bateson's ideas on discontinuous variation and mentioned the *Materials for the Study of Variation* in a chapter entitled "Darwinism attacked" (Kellogg 1907, p. 33). Kellogg considered the selection doctrine as being the true Darwinism and that in the case of discontinuous variation natural selection is absolutely limited in its work to the material furnished by variation (Kellogg 1907, pp. 25, 33).

Robert Olby holds that Bateson started as an orthodox Darwinian, searching for a causal connection between variation and environment (Olby 1955, p. 133). According to William Coleman, Bateson's *Materials for the Study of Variation* represented a return to Darwinism (Coleman 1968, p. 338). Coleman regards Bateson's approach (collecting and codifying facts of organic variation, keeping correspondence with plant and animal fanciers, doing extensive field work, engaging in a comprehensive search of the literature and modest attempts at experimental breeding) remarkably similar to Darwin's.

Alfred Nordmann admits that both Bateson and the biometricians adopted the *Darwinian* research program (Nordmann 1992). Nordmann classifies Bateson as Darwinian because he embraced evolution by natural selection and studied the causes of variation within a broadly Darwinian framework (Nordmann 1992, p. 53).

Peter Bowler suggests that as Bateson gave up the morphological tradition he turned against Darwinism (Bowler 1990, p. 215) and that the *Materials* was his major attack on Darwinism. According to this author, this attack seemed to imply that anatomists and embryologists had, uncritically, assumed all changes to be adaptive (Bowler 1989, p. 284, Bowler 1992, p. 191). He also adds that perhaps the *Materials* had not sold well because it was part of the anti-Darwinian movement (Bowler 1992, p. 191).

In most of those points Bowler does not elucidate what he means by 'Darwinism', why he is labelling Bateson as anti-Darwinian, or why he regards the *Materials* as an attack against Darwinism. However, from general considerations scattered in Bowler's books it is possible to get a better idea of his views towards such issues. He states that Bateson was a strong opponent of selection and the principle of utility; and that he was also indifferent to the Darwinian's efforts to postulate an adaptive purpose for the divergence of form. In addittion, according to him Bateson thought that the discontinuity was the real source of evolution, in opposition to Darwin's gradualness. Besides that, Bowler declares that Bateson advocated orthogenesis.

This short account discloses that there is no general agreement as to Bateson's attitude towards Darwinism. Applying labels to scientists, besides being very hazardous, is not an easy task. In order to avoid meaningless analyses, if one aims to classify X as a Y, it is necessary to state exactly what Y means (or what it is assumed to mean), and to present evidence to the effect that X really does belong to that group. This paper will therefore first discuss the very meaning of "Darwinism" and "Darwinist"; next, it will describe Bateson's early evolutionary work; and then it will attempt to draw a clear-cut conclusion concerning Bateson's attitude towards Darwinism. We think that the use of clear categories may help the historiographical research.

3. What is Darwinism?

There was a time when to many readers Darwinism was synonymous with organic evolution or the theory of descent. According to Peter Bowler, this happened around 1870, when Darwin's theory was the only available explanation of organic change (Bowler 1989, p. 188). Shortly afterwards, however, there arose a wide range of evolutionary proposals. Therefore we should not regard Darwinism as synonymous of organic evolution or theory of common descent, since these features are also common to other theories of evolution.

What specifically characterises Darwinism among evolution theories? 'Darwinism' is a word that is often employed by historians and philosophers of science in the past or today with different meanings which frequently are not clear. Peter Bowler thinks that the theory of natural selection was the core of Darwin's theory and that 'true Darwinism' was based on biogeography and the study of adaptative evolution (Bowler 1990, pp. 14, 140-141, 150). Ernst Mayr holds that in the period immediately after 1859, "Darwinism referred most often to the totality of Darwin's thinking, while it strictly means natural selection for the evolutionary biologist of today" (Mayr 1982, p. 505). Several authors of the late 19th and early 20th century also held that the theory of natural selection was the same thing as Darwinism (Kellogg 1907, pp. 17, 26). The so-called 'neo-Darwinians' of that time would agree with that definition, but other authors who would also like to be called 'Darwinians' (Herbert Spencer, George Romanes, etc.) would not agree. Towards the end of the 19th century, each evolutionist had a different view on what 'Darwinism' meant.

What, after all, should be called 'Darwinism'?

Let us impose at the outset some conditions that, in our view, a suitable delimitation of 'Darwinism' should obey:

1. An adequate concept of 'Darwinism' should include Darwin's own work as 'Darwinian' – otherwise, another name should be chosen, instead of 'Darwinism'.
2. An adequate concept of 'Darwinism' should not exclude everyone else from 'Darwinism' – otherwise, the name would not characterise a research tradition, but the research of a single person.
3. An adequate concept of 'Darwinism' should not include all evolutionists – 'Darwinism' should be regarded as a special type of evolutionary view.

Suppose we deemed that being a Darwinian means *to accept all the features of Darwin's theory*. In that case, it would be very difficult to point out a single Darwinian besides Darwin himself. Thomas Huxley thought that large discontinuous variations were required to understand evolution (Huxley 1860, p. 224); August Weismann could not be included in such category because in his mature work he denied the inheritance of acquired characteristics (Weismann 1904, Vol. 1, chap. 23) that was accepted by Darwin; Alfred Wallace believed that the origin of man was due to supernatural influences, because the human mental capacities could not be explained by natural causes; and so on. Upon close examination, it would turn out impossible to find a single naturalist (or biologist) who followed all the features of

Darwin's original theory. If we take into account those facts, it is very difficult to imagine that Darwin himself might have considered all components of his evolutionary theory as a single indivisible whole as it was suggested by some historians of science (see Mayr 1982, p. 505 for instance). Notice also that this would conflict with requirement (2) above.

In the late 19th century it is clearly observable that many researchers who expressed their sympathy for Darwin's work and accepted many of its features did so, whilst not agreeing with one another. This strongly suggests the existence of a broad Darwinian research program with several conflicting branches. For this reason we will regard as a Darwinian any individual that accepted most of the main features of Darwin's original theory, as elucidated below. Moreover, the naturalists that followed Darwin's methodology, could also be considered in some way as being Darwinians.

4. Darwin's original proposal

Darwin's contribution included several complementary lines of work. He tried to provide evidence for evolution as a fact; he attempted to explain the causes of organic evolution; he studied the role of those causes in individual cases; he tried to reply to objections made referring to his ideas; he pointed out promising lines of research; and so on.

We should begin by reviewing Darwin's original theory of evolution. It is well known that Darwin proposed a theory of descent in several of his works from 1858 onward, trying to explain the origin of different kinds of life (he did not try to explain the origin of life itself). He claimed that living species (including human beings) were not created as they are now, but descended from now extinct species and from common ancestors that were modified by natural causes. One of those natural causes was natural selection − a process similar to the artificial selection consciously or unconsciously used by animal breeders and plant fanciers to produce new varieties and breeds. A natural population is not homogeneous, but composed of many slightly different individuals. Darwin conceived that natural selection performed on slight, continuous variations that were formed according to the law of chance. Such variations were transmitted to the offspring. Not every individual that is born is able to survive and to produce descendants, because there is a limit (food and space limitations) to the increase of living beings. There is a struggle for existence, and those individuals who have a slight advantage over their companions will have a higher probability of surviving, and producing offspring. Since those useful features are hereditary, they will be transmitted to the offspring and this will lead to a gradual modification of the population. The species would be transformed slowly and gradually. Darwin also admitted the existence of sudden (discontinuous) variations whether in animals, cultivated plants or in man, but he did not deem this rare phenomenon as relevant to the evolutionary process. Natural selection is only able to explain adaptive ("useful") features. Darwin also admitted other natural causes, such as sexual selection (to account for "beauty" and secondary sexual features), the direct action of the environment, and the inheritance of acquired characteristics obtained by use and disuse. He also proposed a mechanism for the transmission of such characteristics (the hypothesis of pangenesis). These are in short what could be deemed the main, but not exclusive, aspects of Darwin's theory of evolution.

Darwin suggested as topics for further research the study of domesticated animals and cultivated plants, the mutual affinities of organic beings, their embryological relations, and their geographical distribution as well as the geological succession (Darwin 1872, p. 2).

It is very difficult to perceive which aspects mentioned above could be considered by Darwin himself, in his own time, as the central, unchangeable parts of his work. None of the researchers who are usually labelled as 'Darwinians' (or who called themselves this) have followed all the features of Darwin's original theory. Most of them accepted several characteristics of Darwin's theory, denied others, and sometimes added new ideas. In our view, Darwin inaugurated a research program[1] being broadly conceived and which was also open to new contributions. Or, in other words, as Philip Kitcher suggests, by introducing the Darwinian schemata one recognizes further phenomena about which questions must subsequently be raised (Kitcher 1985, p. 151). 'Darwinism' could fit the ideas or theories proposed

---

[1] We are not advocating or adopting Imre Lakatos' theory but making use of his terminology since we think it is adequate for the discussion of this case.

after Darwin's original theory that had had most of the fore mentioned main features. Furthermore, we think that Darwinism should not be characterized only by a single feature of Darwin's theory, such as natural selection, because Darwin himself stressed that natural selection was the main but not the only cause of transformation of species (Darwin 1872, p. 421).

But, on the other hand, what does it mean to be anti-Darwinian? Ideas or theories proposed after Darwin's that denied all or most of the features of Darwin's original theory could be considered as being anti-Darwinian. In addition to this, scholars that clearly spoke against Darwin's work or methodology could also be deemed as being anti-Darwinians. One for instance is Thomas Hunt Morgan, in his book *Evolution and adaptation* (1903).

5. Darwin's research program

Let us try to make explicit what could be described as Darwin's research program. Specifically, what could be called its 'hard core'? We will adopt a characterization similar to the analysis presented by James G. Lennox (1992).

The 'hard core' of Darwin's research program was:
1. The struggle for survival: Biological organisms have more offspring than can possibly survive.
2. Inheritability: Biological organisms inherit most of their traits from their ancestors and pass them on to their descendants.
3. Variation: The inheritable traits of biological organisms vary, even within the same species (or variety).
4. Differential fitness: Some inheritable traits will be more advantageous than others in the struggle for survival .
5. Natural selection: Therefore, there has been and will continue to be, on average, a (natural) selection of those organisms having advantageous traits that will lead to the evolution of species.

Let us add that natural selection should be regarded here as an important or as the main cause of transformation, but not the *exclusive* cause, according to Darwin.

Darwin's theory predicts (or postdicts) evolution and attempts to explain its cause. However, the theory, by itself, does not say which traits are inheritable, nor how they vary, or the way in which resources are limited, or how different traits aids in survival, or how these factors change over time. Those features, which should be studied in each specific case, make up the 'protective belt' of the theory (Lennox 1992).

According to Lennox, when a particular episode or instance of evolutionary change is investigated, within the Darwinian program, it will be necessary to add specific details to the hard core assumptions:
a. The range of inheritable traits in a biological population (s).
b. The environment, and how it changes over time.
c. The relative benefit that these traits confer to the members of the populations possessing them in the various environments (fitness values). (Lennox 1992)

As natural selection is regarded as the main factor in evolution, one should always first attempt to use it to explain all known characteristics of living beings. Whenever some special cases resist being explained by it, other explanations should be attempted: use-disuse, sexual selection, or other new auxiliary assumptions could be introduced. This could be regarded as the 'positive heuristics' of the Darwinian research program.

6. Bateson's early evolutionary work

William Bateson's scientific career began under the influence of Francis Maitland Balfour. This eminent embryologist, strongly influenced by Darwin, introduced recapitulation tendencies into Cambridge embryology. Like Darwin he also believed that embryological homologies provided good evidence for evolution (see Ridley 1986, pp. 39-40). Bateson, together with his colleagues such as Adam Sedgwick, Walter Frank Raphael Weldon and A. E. Shipley, began his scientific research as an embryologist working in the morphological tradition. Following this approach he attempted to reconstruct the phylogeny of the vertebrates.

Darwin regarded embryonic (excluding larval) characters as being of highest value for the genealogical study of species and their classification (Darwin 1872, p. 368). According to Darwin, Haeckel's work (*General Morphology*) could be regarded as being a great beginning and could show how classification would be treated in the future (Darwin 1872, p. 381). However, Darwin warned that morphology was a much more complex subject than it had first appeared (Darwin 1872, p. 385).

After Balfour's death, Bateson travelled to the Marine Station of Hampton (Virginia) in order to study the development of *Balanoglossus*, under the supervision of William Keith Brooks. Bateson published four papers on this subject, suggesting that the Enteropneusta could be the ancestral of the Chordates. During this study, Bateson's attention was called to the importance of segmentation and metamerism.

Although Bateson's work on *Balanoglossus* was appreciated by the scientific community, he soon gave up the embryological method as a suitable one which searches for phylogenetic relationships.

Bateson pointed out some difficulties related to the embryological method (von Baer's or Haeckel's principle: "ontogeny recapitulates phylogeny"). According to him, despite the magnificent body of facts offered by embryology, the interpretation of such facts was still to be sought after (Bateson 1894, p. 10). The interpretation of embryological evidence demanded some hypotheses such as the course of variation in the past. Depending on the chosen hypothesis one could arrive to opposite conclusions: "The Embryological Method then has failed not for want of knowledge of visible facts of development but through the ignorance of the principles of Evolution" (Bateson 1894, pp. 5, 9). However, as Alan Cock points out (Cock 1973, pp. 13-14), Bateson did not deny that embryology provides some guidance to the discovery of phylogenetic origins, nor that adaptation is a genuine phenomenon and a proper subject of study (Bateson 1894, pp. 7-13).

In a letter to Adam Sedgwick Bateson emphasized his view: "About embryology [...] we shall talk someday. But I am sorry that I now seem to hold the facts of embryology lightly – my quarrel is only with the precision of interpretation" (Letter from William Bateson to Adam Sedgwick, 5 February 1894. Cambridge University Library, Mss. Add. 8634, A.9.a.4).

Should Bateson's attitude as regards embryological analysis be considered as an attack on Darwinism? We don't think so. Darwin never undertook an embryological research – he only suggested that it would be useful. However, the possibility of ascertaining with certainty the genealogical relationship between organisms by embryological analysis should not be regarded as an essential feature of Darwinism. Bateson started from a Darwinian line of research, undertook a specific embryological study, and faced all the problems concerning the reconstruction of phylogenies. As he concluded that the method did not lead to certainty, he moved to another approach that was also compatible with Darwinism.

From 1886 to 1887 Bateson spent 18 months in the Siberian Steppe studying the environmental influences on the variation of the salty lakes fauna. However, such studies also remained inconclusive since he could not decide whether the observed modifications were due to physical changes (a kind of effect accepted by Darwin) or produced by natural selection.

Since his youth Bateson accepted the principle of the struggle for life. In a letter to his mother, written during his studies in the Steppe, he remarked: "Life without killing and without a struggle cannot go on. It is impossible to diminish the intensity of the struggle [...]" (Letter from William Bateson to his mother, 19 June 1887 – 11 July, Cambridge University Library, Mss. Add. 8634, G1b).

7. Bateson's attitude in the *Materials for the Study of Variation*

After returning from the Steppe, Bateson devoted himself for seven years to the collection of facts relating to variation which culminated in the publication of his book *Materials for the Study of Variation* (1894), a catalogue of facts that substantiated the discontinuity of variation.

Bateson was interested in the problem of the origin of species (Bateson 1894, pp. xi, 571) and intended to undertake a "serious study of variation and to make sure a base for the attack on the problems of Evolution" since he thought that this problem had "remained unsolved and that old questions standed unanswered" (Bateson 1894, pp. xi-xii). He stated that "To collect and codify facts is the first duty of the naturalist" (Bateson 1894, p. vi). Here he was clearly inspired by Darwin. In his work *The Variation of Animals and Plants Under Domestication*, Darwin had already collected some very good facts relevant to the study of evolution. Bateson attempted to do a similar job for wild species.

In the 'Introduction' of his book Bateson respectfully wrote about Darwin's role and importance in the study of evolution:

> It is more than thirty years since the *Origin of species* was written, but for many these questions are in no sense answered yet. In owning that it is so, we shall not honour Darwin's memory the less; for whatever may be the part which shall be finally assigned to Natural selection, it will always be remembered that it was through Darwin's work that men saw for the first time that the problem is one which man may reasonably hope to solve. If Darwin did not solve the problem himself, he first gave us the hope of a solution, perhaps a greater thing. How great a feat this was, we who have heard it all from childhood can scarcely know (Bateson 1894, p. 1).

Bateson was doubtlessly an evolutionist: he accepted the doctrine of descent. "In what follows, it will be assumed that this Doctrine of Descent is true". He adopted it as a postulate and considered it as a fundamental conception in evolution (Bateson 1894, pp. 4, 14).

Next, Bateson discussed the two main *explanations* of evolution: Lamarck's and Darwin's theories. Bateson described the main points of Darwin's theory in the following way:

> Darwin, without suggesting causes of Variation, points out that since (1) Variations occur – which they are known to do – and since (2) some of the variations are in the direction of adaptation and others are not – which is a necessity – it will result from the conditions of the Struggle for Existence that those better adapted will *on the whole* persist and the less adapted will on the whole be lost. In the result, therefore, there will be a diversity of forms, *more or less* adapted to the states in which they are placed, and this is very much the observed condition of living things (Bateson 1894, p. 5).

Notice that Bateson emphasized that he strongly agreed with (1) and (2) and with their consequences.

Did Bateson accept that natural selection was a correct explanation? Yes, he acknowledged Darwin's views as providing a true explanation (*vera causa*) of the transformation of living beings:

> It may be remarked however that the observed cases of adaptation occurring in the way demanded on Lamarck's theory are very few, and as time goes on this deficiency of facts begins to be significant. Natural Selection on the other hand is obviously a 'true cause' at the least (Bateson 1894, p. 5).

Although Bateson did not regard Lamarckism as a viable alternative, he sent a copy of his book to the French Neo-Lamarckian evolutionist Alfred Giard. After receiving Bateson's *Materials for the Study of Variation*, Giard wrote:

> Vous ne me croiriez pas si je vous dirais que nous sommes d'accord sur tous les points. Si peu que vous ayez fait de theorie vous en avez fait assez (et je vous en félicite) pour qui nous ayons le droit de batailler un peu et bien que vous ne soyez pas a la manière de Weismann plus Darwiniste que Darwin je ne vous trouve pas toujours assez Lamarckien. Cela n'empeche pas que vous avez fait une oeuvre excelente et d'une utilité incontestable (Letter from Alfred Giard to William Bateson, 15/2/1894, Cambridge University Library Mss. Add. 8634, A.9.a.3).

Giard clearly perceived that Bateson was neither a Lamarckist nor a follower of the so-called Neo-Darwinian school. Although Bateson accepted the theory of natural selection as proposed by Darwin in the *Origin*, he did not agree with some of its later representations:

> In the view of the phenomena of Variation here outlined, **there is nothing which is in any way opposed** to the theory of the origin of Species "by means of Natural Selection or the preservation on favoured races in the struggle of life". But by **a full and unwavering belief in the doctrine as originally expressed**, we shall in no way be committtted to the representations of that doctrine made by those who have come after (Bateson 1894, p. 80; our stress).

Notice how clearly Bateson stated here his belief in natural selection. If Bateson was not lying when he said he accepted Darwin's theory and of course he was not, what was he attempting to do? It seems that he was trying to improve the theory and to account for some difficulties that had not been answered.

8. Bateson's criticism of former methods

According to Bateson, one of the chief difficulties of Darwin's theory (and Lamarck's) was to explain why species exhibit a discontinuous series, while the physical environment presents a continuous gradation of temperature, altitude, depth of water, salinity, etc. (Bateson 1894, p. 5). Darwin was aware of this difficulty and attempted to provide an answer, but Bateson could not accept it. He regarded this problem as a fatal objection to the supposition that all variation was continuous and the discontinuity of species resulted from natural selection (Bateson 1894, p. 69). Therefore he attempted to find elsewhere the source of the discontinuity of species – to wit, in the very process of variation.

Before entering into the subject of variation, however, Bateson attempted to show other approaches to the study of evolution principally that the embryological method and the study of adaptation had failed: "It is besides in the examination of these methods and in observing the exact point at which they have failed, that the need for the Study of Variation will become most evident" (Bateson 1894, p. 5). In those criticisms some authors have found evidence that Bateson was attacking Darwinism.

Bateson's comments on the embryological method have already been presented above. Let us now present his criticism on the study of adaptation.

According to Darwin's theory, natural selection can explain useful features of living beings. Many complex structures were however found in many classes of animals which had no known use to those animals (Bateson 1894, p. 10). Besides that, many of the differences between related species seemed useless and trivial. He also made objections to the study of adaptation as means of discovering the processes of evolution because while it is possible to suggest some way by which in circumstances, known or hypothetical, any given structure may be of use to any animal, it cannot on the other hand ever be possible to prove that such structures are not on the whole harmful either in a way indicated or otherwise (Bateson 1894, p. 12).

To sum up, both in the case of the embryological method as in the case of the study of adaptation, Bateson thought that it was possible to *suggest* but it was impossible to *establish* how organisms had evolved in specific cases.

Could this criticism be regarded as an attack against Darwinism? No. The difficulty or even impossibility of finding out in individual cases how living beings evolved does not amount to denying they evolved according to Darwin's theory. Bateson was not proposing an *alternative* to Darwin's theory. Indeed, the method he supported was not intended to find out the genealogy of species or the origin of specific features. Bateson's foremost aim in studying variation was this:

> The first question which the Study of Variation may be expected to answer, relates to the origin of that Discontinuity of which Species is

the objective expression. Such Discontinuity is not in the environment; may it not, then, be in the living thing itself? (Bateson 1894, p. 17).

Neither embryological investigations nor the study of adaptations had this aim in view. Besides that, those former methods were not incompatible with the study of variation. It seems that Bateson's criticism of other lines of research only intended to show that they were not as valuable as they were supposed to be, and that other kinds of investigation merited attention.

It is also relevant to point out that both the first (1859) and the 6th edition (1872) of the *Origin* were part of Bateson's private library.[2] The fact that he marked out some paragraphs of both editions suggests that he deemed them relevant in some way. In the 1859 edition, for instance, Bateson marked out a paragraph in chapter IV where Darwin mentioned some peculiarities that were observed to arise and to become attached to the male sex in domestic animals that seemed of no use to the males in battle nor attractive to the females. Darwin also mentioned analogous cases that could be found under nature (Darwin 1859, p. 90). This strongly suggests that this point attracted Bateson's attention and that the difficulties he noticed in the study of adaptation arose from his study of Darwin's works.

9. Discontinuous variations

Bateson thought that the study of variation – that is, the study of differences between organisms[3] – could throw some light on the problem of evolution. Such study might begin by the determination of the nature of the series by which different forms evolved. How had differenciation been introduced in these series? To know if these series were continuous (if the transition from term to term is minimal and imperceptible) or discontinuous (when there are gaps filled by no transitional form) and to decide which of them agrees most with the observed phenomena of variation was a vital question concerning the study of evolution. According to Bateson this question had never

---

[2] Bateson's books are now in the possession of the John Innes Centre in Norwich, England.
[3] The organisms that are compared must be parent and offspring. If the actual parent is unknown, the normal form of species must be known (Bateson 1894, p. 17).

been decided. However, it was commonly accepted that the process was a continuous one (Bateson 1894, pp. 14-15).

Bateson admitted that in many cases variation could be continuous:[4]

> The fact that Continuous Variation exists is also none the less a fact, but it is most important that the two classes of phenomena should be recognized as distinct, for there is a reason to think that they are distinct essentially, and that though both may occur simultaneously and in conjunction, yet they are manifestations of distinct processes. The attempt to distinguish these two kinds of Variation from each other constitutes one of the chief parts of the study (Bateson 1894, p. 18).

It is important to stress that Bateson never denied the existence of continuous variation. His position towards this issue was quite clear. Three years after the publication of the *Materials* he wrote:

> Whether continuity or discontinuity is found depends on the species studied and the character selected for investigation. There is continuous variation, but there is discontinuous variation also. To discover by statistical investigation the degree of continuity or discontinuity which in each species is manifested by the variation of character is the first business of the student of evolution (Bateson 1897, pp. 346-347).

In the *Materials* Bateson accumulated a large number of cases (886) that substantiated the existence of discontinuous variations. He mainly dealt with the variations that took place in meristic[5] series (radial or linear). Such a kind of variation affected the many different parts of organisms.

It is highly probable that Bateson was inspired by Darwin in this decision. In Bateson's exemplar of Darwin's book, *The Variation of Animals and Plants Under Domestication* which was part of his private library, Bateson added a question mark [?] beside the following text:

---

[4] He mentioned for instance some cases studied by Francis Galton and Raphael Weldon (see Bateson & Bateson 1891, p. 158).

[5] Bateson described merism as a phenomenon of repetition of parts (symmetry or pattern) (Bateson 1894, p. 20). This author also admitted the existence of another kind of variation which he called 'substantive' variation. This phenomenon encompasses variations in the constitution of the parts themselves (Bateson 1894, p. 23). Although realizing that some substantive variations behaved discontinuously, Bateson did not name them in his text of *Materials*.

"I allude to organs which are abnormally multiplied or transported. Thus the gold-fishes often have supernumerary fins placed on various parts of their bodies" (Darwin 1998, p. 398). Certainly such a statement left him not only puzzled but also curious. Of course, this represented a problem to Darwin's belief that the gradual accumulation of very slight variations by natural selection was the mechanism of evolution, and that aberrant types had no evolutionary importance.

Although Bateson did not deny natural selection, he noticed that the theory presented several problems that were very difficult to answer if one assumed that variations were always very small (continuous). He wrote about what he regarded as the most serious objection to the theory concerning the building up of new organs in their initial and imperfect stages, the mode of transformation of organs, and generally, the selection, perpetuation, and utility of minute variations:

> [...] Assuming that variations are minute, we are met by this familiar difficulty.[6] We know that certain devices and mechanisms are useful to their possessors; but from our knowledge of Natural History we are led to think that their usefulness is consequent on the degree of perfection in which they exist, and if they were all imperfect, they would not be useful. Now it is clear that in any continuous process of Evolution such stages of imperfection must occur, and the objection has been raised that Natural Selection cannot protect such imperfect mechanisms as to lift them to perfection. Of the objections which have been brought against the Theory of Natural Selection this is by far the most serious (Bateson 1894, pp. 15-16).

In the case of discontinuous variation, the objection loses its strength, of course. Notice that, in this case, the idea of discontinuous variation is intended to *help* the theory of natural selection, and not to replace it.

Concerning the utility and perpetuation of minute variations, Bateson gave the example of many South African butterflies of the genus *Euchloe*. Some of them have the apices or tips of the fore-wings orange-red (*Euchloe danae*), while in others they are purple (*Euchloe ione*). He stated that if it is assumed that the transition from orange to purple and vice-versa is continuously affected by successive selection

---

[6] Such an objection had been already raised in Darwin's time by St. George Mivart: "Natural selection is incompetent to account for the incipient stages of useful structures". Darwin discussed it in chapter 7 of *The Origin of Species* (Darwin 1872, p. 177). See also Kellogg (1907, p. 49).

of minute variations some difficulties would arise. "Why is purple a good colour for this creature? If purple is a good colour and red is a good colour, how did it happen that at some time or other all the intermediate shades were also good enough to have been selected? and so on" (Bateson 1894, p. 72). Bateson presented another possibility: "I submit that it is easier to suppose that the change from red to purple was from the first complete, and that the choice offered to Selection was between red and purple; and the tints of the purple and of the red were determined by the chemical properties of the body to which the colour is due" (Bateson 1894, p. 73). This problem had been also discussed in a previous work of Bateson (Bateson & Bateson 1891, p. 128).

In many cases where the number of parts of a specific organ in a population was observed to vary, Bateson noticed that the supernumerary parts were perfect, and that the organ with an anomalous number of parts retained its symmetry. Those were examples of discontinuous variations that provided perfect and symmetrical organs. Bateson held that it was impossible to suppose that the perfection of a variety,[7] discontinuously and occurring suddenly, was produced by selection. He clarified it with an example referring to a variation in the number of petals in a variety of tulips:

> No doubt it is conceivable that a race of Tulips having their floral parts in multiples of four might be raised by Selection from a specimen having this character, but it is not possible that the perfection of the nascent variety can have been gradually built up by Selection, for it is, in its very beginning, perfect and symmetrical. And if it may be seen thus clearly that perfection and Symmetry of a variety is not the work of Selection, this fact raises a serious doubt that perhaps the similar perfection and Symmetry of the type did not owe its origin to Selection either. This consideration of course touches only the part that Selection may have played in the first building up of the type and does not affect the view that the perpetuation of the type once constituted, may have been achieved by Selection (Bateson 1894, p. 69).

---

[7] Bateson admitted that the perfection and definiteness of the type could be due to the physical conditions under which variation proceeds. In this way, he formulated not a theory but a working hypothesis that the discontinuity of meristic variation (found in *Tulip*, *Aurelia* or the cockroach tarsus) could be determined *mechanically* (Bateson 1894, pp. 69-70).

In short, Bateson did not reject natural selection. He thought that natural selection could not create new features but, once those features arose (mainly by discontinuous variation), it would act in the perpetuation of the type.[8] Some years later he stated in an address to the British Association for the Advancement of Science: "Selection is a true phenomenon; but its function is to *select*, not to create." (Bateson 1904, p. 238)

Now, Darwin assumed that natural selection acted upon slight, continuous variations. Could Bateson's work be considered anti-Darwinian, for that reason?

Before answering this question, let us remark that Darwin had never been acquainted with the facts brought to the light through Bateson's work. Of course, Darwin became familiar with wild plants and animals in his Beagle voyage, but at that time he was not concerned with evolution and did not collect information concerning variation. When he turned to variation, he concentrated his attention on domestic varieties, believing that variation was much smaller in wild plants and animals: "When we compare the individuals of the same variety or sub-variety of our older cultivated plants and animals, one of the first points strikes is that, they *generally* differ more from each other than do the individuals of any one species or variety in a state of nature." (Darwin 1859, p. 9; our emphasis)

Bateson, however, found evidence of the effect that wild animals could have as changing as domesticated ones can be. After examining the variation in the vertebrae of the sloths, in the teeth of the anthropoid apes as well as in the colour of the dog-whelks (*Purpura lapillus*) he noticed a frequency and a range of variation matched only by the most variable of domesticated animals (Bateson 1894, p. 572). Besides that, Bateson was able to show that discontinuous variation was very common, a fact that was also unknown to Darwin.

---

[8] The same idea may be found in a previous article of Bateson and his sister Anna about the variation in floral symmetry of certain plants having irregular corollas. The authors stated that "the evolution of the forms of the irregular corollas had occurred in connection with their adaptation to the purpose of cross-fertilisation, and their perfection and persistence have consequently been achieved by the agency of Natural Selection." (Bateson & Bateson 1891, p. 126)

It seems that the hypothesis of continuous variation was used by Darwin only because he thought that this was a nice account of observed facts. The opposite possibility of discontinuous variation is compatible with all the other features in Darwin's theory, and in addition presents several advantages upon the hypothesis of continuous variation. Had Darwin known that discontinuous variations are common among wild plants and animals, why would he refuse to use this fact to answer to known difficulties of his theory?

The emphasis upon discontinuous variation and its role in evolution was a significant new step. It went against one of Darwin's secondary assumptions. However, this step should not be deemed as an attack against Darwinism. It was an attempt to improve Darwin's theory.

10. Conclusion

Bateson accepted as a postulate that evolution had occurred, therefore he was an evolutionist. It is also clear that he did not support Lamarckian theories. So was he a Darwinian, after all?

If we state that Darwinism is synonymous with the theory of natural selection and that natural selection is the only and exclusive way of modification, only authors such as August Weismann (in his mature work) could be classified as Darwinians: "It seems difficult to refuse to admit [...] *that every essential part of a species is not merely regulated by natural selection, but it is originally produced by it*" (Weismann 1904, Vol. 2, p. 312). Using such a criterion, Bateson could not be regarded as a Darwinian. But if we assume Weismann's view, even Darwin could not be considered a Darwinian, because in the 6th edition of *On the Origin of Species* he repeated something that he had been stating since 1859: "I am convinced that natural selection is the main, although not exclusive way of modification." (Darwin 1872, p. 421)

Bateson not only admired but also regarded Darwin's work as very important to the study of evolution. He wrote respectfully about Darwin and accepted the theory of natural descent. He was also influenced by Darwin methodology.

If we compare Bateson's early evolutionary work and the *hard core* of Darwin's theory as described above, we notice no conflict. Bateson accepted the struggle for survival, inheritability, variation, differential fitness, and natural selection. Although he pointed out some difficulties of the theory, he did not deny it.

Concerning the protective belt, since Darwin's theory itself did not state which traits are inheritable, nor how they vary, or the way in which resources are limited, or how different traits aid in survival, or how these factors change over time, Bateson's contributions and suggestions are plainly compatible with the theory.

If we take into account other criteria adopted by the authors to label Bateson as an anti-Darwinian we will notice that they are inadequate. It was shown, for instance, that Bateson was not an anti-selectionist, as claimed by a few authors. Although emphasizing the relevance of discontinuous variation, he did not deny the existence of continuous variation. However, he accumulated a great mass of facts (in the Darwinian spirit) that suggested that discontinuous variation was not as rare as was thought, and drew some conclusions regarding evolution.

As suggested by Nordmann, Bateson worked in a broad Darwinian framework of problems and questions (Nordmann 1992, p. 53). He was, as most of his colleagues, deeply influenced and inspired by Darwin's work. Bateson's approach in the *Materials* was remarkably similar to Darwin's, as Coleman pointed out (Coleman 1968, p. 338). Bateson not only followed, but also referred to Darwin's achievement respectfully.

Some authors that classify Bateson as anti-Darwinian regarded Wallace as being a Darwinian since he was converted to evolutionism by the evidence of biogeography. He is also described as a strong proponent of natural selection (Bowler 1990, pp. 140-41). However, it should be recalled that Wallace did not accept sexual selection, and claimed that men are not like other animals since they are endowed with a reason that was not developed by natural selection or any other natural cause. *This* strongly conflicts with the hard core of Darwinism. There are better reasons to exclude Wallace from the Darwinian research program than to classify Bateson as an anti-Darwinist.

Whereas Darwin did not regard sudden and large variation as relevant to the processes of evolution and presented facts bearing on the continuity of variation, Bateson accumulated a huge mass of facts concerning the discontinuity of variation. However, he did not deny the existence or the importance of continuous variation.

Bateson increased the collection of facts known to Darwin. The fact that many forms are perfect and arise suddenly, represented a problem concerning the action of natural selection. How could it act in the production of new forms if they were perfect from the beginning? In spite of this and other restrictions to natural selection, Bateson accepted that it acted in the perpetuation of the type.

This study leads to the conclusion that Bateson's *Materials* cannot be regarded as an attack against Darwinism, but as a contribution to the Darwinian research program. It contributed to Darwin's theory, adding new facts, suggesting new features of evolution and providing an original solution to certain difficulties in the theory. Bateson's work in his earlier professional career may be regarded as part of a broad Darwinian research programme with several conflicting branches.

References

Bateson, B. (1928), *William Bateson, F.R.S. Naturalist. His essays and addresses, together with a short account of his life*, Cambridge: Cambridge University Press.

Bateson, W. (1894), *Materials for the Study of Variation*, Baltimore: Johns Hopkins, 1992.

Bateson, W. (1894), "Progress in the Study of Variation, I", *Science Progress* 1, 1897, reissued in Punnett (1928), pp. 344-56.

Bateson, W. (1904), "Presidential Address to the Zoological Section, British Association, Cambridge Meeting, 1904", reissued in Bateson (1928), p. 233-259.

Bateson, W. & A. Bateson (1891), "On the Variations in Floral Symmetry of Certain Plants Having Irregular Corollas", *Journal of the Linnean Society* (Bot.) 28, reissued in Punnett (1928), pp. 126-161.

Bowler, P. (1989), *Evolution: The History of An Idea*, 2nd ed., Berkeley: University of California Press.

Bowler, P. (1989), "Development and Adaptation: Evolutionary Concepts in British Morphology, 1870-1914", *British Journal for the History of Science* 22: 283-97.

Bowler, P. (1990), *Charles Darwin. The Man and His Influence*, Cambridge: Cambridge University Press.

Bowler, P. (1992a), *The Eclipse of Darwinism: Anti-Darwinian Evolution Theories in the Decades Around 1900*, 2nd ed., Baltimore: Johns Hopkins University.

Bowler, P. (1992b), "Foreword", in Bateson (1894[1992]), pp. xvii-xxvii.

Cock, A. (1973), "William Bateson, Mendelism and Biometry", *Journal of the History of Biology* 6: 1-36.

Coleman, W. (1968), " On Bateson Motives for Studying Variation ", *Actes du XIème Congrès International d'Histoire des Sciences*, Varsovie–Cracovie, 24-31 Août, 1965, 6 vols., Ossolineum: Académie Polonaise des Sciences, 1968, Vol. 5, pp. 335-339.

Darwin, C. (1859), *On the Origin of Species by Means of Natural Selection or The Preservation of Favoured Races in the Struggle of Life*, London: John Murray.

Darwin, C. (1872), *On the Origin of Species by Means of Natural Selection or The Preservation of Favoured Races in the Struggle of Life*, 6th ed., London: John Murray.

Darwin, C. (1888), *On the Origin of Species by Means of Natural Selection or The Preservation of Favoured Races in the Struggle of Life*, 6th ed., with additions and corrections, London: John Murray.

Darwin, C. (1868), *The Variation of Animals and Plants Under Domestication*, 2 vols. London: John Murray.

Darwin, C. (1998), *The Variation of Animals and Plants Under Domestication*, 2 vols. Baltimore and London: Johns Hopkins University.

Kellogg, V.L. (1907), *Darwinism To-Day*, New York: Henry Holt.

Kitcher, P. (1985), "Darwin's Achievement", in Rescher, N. (ed.), *Reason and Rationality in Natural Science*, Lanham, MD: University Press of America, pp. 123-185.

Lakatos, I. (1970), "Falsification and the Methodology of Scientific Research Programmes", in Lakatos, I. & A. Musgrave (eds.), *Criticism and the Growth of Knowledge, Proceedings of the International Colloquium in the Philosophy of Science*, London, 1965, Vol. 4, Cambridge: Cambridge University Press, pp. 91-195.

Lennox, J.G. (1992), "Philosophy of Biology", in Salmon, M.H. (ed.), *Introduction to the Philosophy of Science*, chapter 7, Englewood Cliffs, N.J.: Prentice Hall, pp. 269-309.

Mayr, E. (1982), *The Growth of Biological Thought. Diversity, Evolution and Inheritance*, Cambridge, MA: Harvard University Press.

Morgan, T.H. (1903), *Evolution and Adaptation*, New York: MacMillan.

Nordmann, A. (1992), "Darwinians at War. Bateson's Place in Histories of Darwinism", *Synthese* 91: 53-72.

Olby, R. (1966), *Origins of Mendelism*, London: Constable.

Punnett, R.C. (ed.) (1928), *Scientific Papers of William Bateson*, Cambridge: Cambridge University Press.

Ridley, M. (1986), "Embryology and Classical Zoology in Great Britain", in Horder, T.J., Witkowski, J.A. & C.C. Wylie (eds.), *The Eight Symposium of the British Society for Developmental Biology. A History of Embryology*, Cambridge: Cambridge University, pp. 35-67.

Weismann, A. (1904), *The Evolution Theory*, 2 vols., London: Edward Arnold; reprinted 1983.

Newton Freire-Maia and Human Genetics in Brazil

Nadir Ferrari

Nucleous of Studies in Human Genetics (NUEG)
Federal University of Santa Catarina (UFSC)

1. Introduction

This is a story about Newton Freire-Maia, considered a pioneer in the formation of the human geneticist's community in Brazil and the first one to publish in this area. Professor Newton has been my teacher, the adviser of my Master's thesis and a dear friend. This friendship and the admiration that I feel towards his work will certainly influence the approach of the narrative. Besides written documents, testimonies of his colleagues, friends and relatives will be used, as well as memories from situations experienced by myself.

In order to understand the trajectory of Freire-Maia, as a scientist of his time, I have been studying the history of the institutions that in one way or another played a role in his professional life. These studies have resulted in two essays: one about the Rockefeller Foundation and the role it played in the development of human genetics in Brazil (Ferrari 1977), and the other about the Universidade Federal do Paraná [Federal University of Paraná] (Ferrari 2000).

Although Freire-Maia is the main focus, other scientists belonging to the first generation of Brazilian human geneticists such as F.M.

Salzano, O. Frota-Pessoa, P.H. Saldanha and Cora de Moura Pedreira are also included.

I first heard about Freire-Maia when I read an article published in a newspaper in 1972. The report said that he and his team were studying the population of a tiny island called Lençóis, near the coast of Maranhão State. The place was considered a Genetic Isolate and presented a high incidence of albinism.

The universe of scientific research represented, then, something very distant, almost unreal, to an undergraduate student in her first years of study, so it would have been natural that my memory of this person faded with time. However, since then my mind has kept adding importance to that first encounter, first because genetics ended up being my main interest and ultimately because I would become one of the many people strongly influenced by Freire-Maia. For this reason, this story will start with the story of this albinism research.

2. The research about genetic drift

Newton first became interested in the community of Ilha dos Lençóis after reading an article published in *O Cruzeiro*, an important magazine at that time. He organized a scientific excursion supported by the World Health Organization and the Rockefeller Foundation. Took part in the expedition: a cytogeneticist, an ophthalmologist, a dermatologist and a general practitioner, for the article said that there were many ill people on the island, besides the cases of albinism. Apart from other errors and imprecision of the article, the name mentioned was of another island, close to the Maiaú archipelago. So, at their arrival in São Luís, the first task was to find out the right name of the place and how to get there. After asking around, Newton discovered that there was one person that could help them, a salt bed landlord who lived in São Luís with his family and owned a house on an island near Ilha dos Lençóis. The researchers had to fly from São Luís to the town of Cururupu, and then take a smaller plane lent by the government in order to land on that island, due to the limitations of the airport there. Once there, they were hosted by the salt bed landlord and, daily, would take a boat to Ilha dos Lençóis. The State governor at that time was an old College friend of Lysandro Santos Lima's at the medical school in Rio de Janeiro. So, the support received by the team was partially motivated by the friendship between Lysandro and Freire-Maia. The salt bed owner became a friend of Newton, and, as a businessman who knew everyone on the island, would go along and introduce the group to the locals, facilitating investigations. A reporter from *Veja* magazine and another from *O Cruzeiro* took the same trip and provided full coverage.

During one of his classes in the genetics graduate course, Professor Newton would describe to us the fascinating family structure of the studied community, to help us understand the meaning of genetic drift. Although his speech would always appraise scientific research and lessen teaching, he enjoyed giving classes and captivated his students. He would talk about discoveries he had made or witnessed, present as his friends the great names we had only seen in genetics books and bring science into our everyday life. On that island, also known as Enchanted Island, women started having children at 15 or 16 years of age and would marry and divorce (marriage by consensus) without difficulties, sometimes with much older or much younger men. Some children would stay with their mother and others with their father, therefore, during the interviews for genealogical identification, stories would come up like: "Ah! First, I will tell you about when I was married to Zequinha. We had these two children you see here, plus another one that Maria is raising, because she married Zequinha after we broke up. My son Pedro is from my marriage with João and the youngest girl there is mine and Luis's, to whom I am married now." Another peculiarity of this population is that, unlike other small isolated communities, it presented a relatively small incidence of consanguineous marriage.

In genetics, a gene is considered deleterious when, in the homozygous state (if it is recessive) or even in the heterozygous state (if it is dominant), it affects negatively the viability and fertility of the individuals that carry it. In other words: it increases precocious mortality and decreases fertility. Consequently, bearers of deleterious genes have a smaller chance to adapt, that is, they leave less descendents and their genes are less likely to be passed on to future generations. Viability and fertility are among the characters that define the adaptive value (Darwinian adaptation). However, on Ilha dos Lençois, it was observed that albinos were not different from normal people concerning early death or fertility. This means that natural selection did not seem to be working against the gene that causes albinism, despite increasing the chances of skin cancer due to the lack of protection against the ultraviolet radiation from sunlight. Freire-Maia's conclusion was that the rising number of albinos on the island was not a consequence of natural selection, which is a directional, deterministic event, but rather due to genetic drift, a non-directional, stochastic event.

It is important to stress that, at that time, most researches about evolution, both in human and non-human populations, focused on systematic pressures rather than random events (Freire-Maia *et al.* 1978), therefore the importance of Freire-Maia's works about Genetics Drift.

3. Brief chronology

Freire-Maia's first scientific paper was published in 1947 and, in 1950, his first essay on human genetics was the tenth on a list that, today, comprises about 500 publications, including notes, essays, books and chapters of books published in Brazil, and other 14 countries. His works comprise mainly the following themes: chromosomal polymorphism in the genus Drosophila, pigmentation polymorphism in the genus Drosophila, consanguineous marriage, congenital malformations, ectodermic dysplasias, precocious morbidity and mortality among sons of radiologists practitioners, migration, normal genetic characters, psychiatric genetics, genetic drift, mathematics genetics, evolution and, more recently, philosophy of science and science and religion. His book on ectodermic dysplasias, written in collaboration with Marta Pinheiro and edited in New York, presents 22 new entities described by him and his collaborators and was the first book ever published on the subject.

4. Scientist *en herbe*

The initiation of Freire-Maia into the scientific life happened long before 1947, but it would be difficult to pinpoint when it took place, because it would also depend on our concept of scientific activity. Newton's first demonstration of love for science occurred when he was in his second year of junior high school. When visiting a friend who had appendicitis, he heard the doctor's explanations about the digestive system and became very curious about it. He started studying the subject in every book he could find, and then started to study all other living beings. From natural history books he turned to biology ones, that contained many genetics subjects, and the idea of becoming a scientist began to rise.

Newton's grandfather, Domiciano Juvêncio Maia, who liked to be called Sô Sano, was what could be called an amateur scientist. Pharmacist at the small town of Boa Esperança in Minas Gerais, he had at the back of his store a vast laboratory, consisting of two big rooms full of scientific instruments and glassware where he performed parasitological and serologic exams. When Newton's father, Sô Sano's only son, graduated as a pharmacist, Sô Sano told him: "Now I will do what I like, I'll study." And he spent all his life studying. Newton knew him like that, studying all day long, always carrying a book, or experimenting in his lab, or observing through his microscope. He enjoyed teaching and would summon Newton and any other kid available to laboratory lessons, at which they would perform simple chemistry experiments and answer questions. Newton tells that Sô Sano never cared about money, he spent his life donating medicine to poor people, and never sent the bill to those who had not paid. His main concerns were science and music. Many times, when people came to him to pay off their debts he would send them away saying "I'm too busy now. Come back some other time... there's no hurry..."

Newton did not pay much attention to his grandfather's lessons, but the lab atmosphere and the image of this unusual grandfather imprinted in him a different way of looking at life and science, that was only evident later and that he acknowledges in the dedication of some of his books. When he saw the X-ray the doctor used to explain his friends appendicitis, he realized that science was applied in everyday life and understood the things his grandfather had explained. Also, Sô Sano may have influenced the socialist ideas that prevailed in Freire Maia's actions rather than in his speeches and were responsible for bitter moments during military dictatorship.

From very early age, Newton had an autonomous and inquisitive mind, resistant to formal education. During his first school years he wrote three novels, several speeches and poetry. Entertained by his father's typing machine, he would forget, sometimes, to go to school. Sometimes with a film projector given by the unusual grandfather, who played the organ, the violin, and the clarinet, he would bind together pieces of old movies to improvise a cinema at home and charge friends for tickets.

After his friend's appendicitis episode, he began to read every biology book he had access to. When his youngest brother was born, already in love with genetics, he wanted him to be called Mendel, but the baby was called José Domiciano.

During High School in Belo Horizonte City, Newton would contest his teacher who did not accept Mendel's laws, and received the nickname "palpiteiro" (nosy). In his memories, published in 1995, he tells: "I lived in other worlds, worrying about journalism, radio, literature, science [...]. Teachers were excellent, colleagues too, but what they tried to teach me was exactly what I didn't care to learn." Newton only wanted to study genetics and philosophy. He would write essays on politics – World War II was starting and he firmly positioned as anti-fascist. He wrote a book called *Heritage and Life*, which his father agreed to sponsor. It had a chapter on life, one on heritage, another about germs – according to Weismann – a chapter on immortality of unicellular organisms and one about the inheritance of acquired characters. In his own words: "Full of errors, a hideous thing." Still, the book was praised in magazines and newspapers of the time. An article published by Jornal do Brasil 06/26/1939, for example, ends with "[...] for what he has accomplished so far it would not be too optimistic to predict to this young man a great achievement in a near future."

After three years in Belo Horizonte without passing the freshmen year exams at high school, and certain that what he wanted was to be a scientist, Newton went looking for work at laboratories in Rio de Janeiro. He enjoyed Carnival, got a girlfriend and lived for two years on the allowance sent by his father. He even applied for a job at a laboratory affiliated with the Ministry of Agriculture, but failed to answer the questions, all concerning zoology and botanic, none about genetics! He managed to find a job at an insurance company, which he managed to bear for three months, and after that he went back to Boa Esperança.

His father convinced him that, to be a scientist, he needed higher education, whichever it was, and Newton agreed to go to the city of Alfenas, where he could finish high school, get into College and at the same time teach, because both the grade-school and College were directed by a family friend.

While living in Alfenas, Newton would often go to Rio de Janeiro for the National Students Union meetings and tried to obtain the General Biology chair at the Natural Sciences College there, ran by A.G. Lagden Cavalcanti and where Frota-Pessoa worked. He did not get the job, but became a friend of Frota-Pessoa and met André Dreyfus, who ran the Biology Department at the Universidade de São Paulo [São Paulo State University].

5. Initiation ritual

In 1943, Theodosius Dobzhansky came to Brazil for three months. This Russian researcher, naturalized American, was one of the leading scientists to formulate the Synthetic Theory of Evolution, which was the result of the population approach of the theory of natural selection and of mendelian genetics, with the contribution of various disciplines. The first edition of Dobzhansky book *Genetics and the Origin of Species* in 1937 called the attention to other factors, besides gene mutations, to be considered in the theory of evolution. He formulated his ideas mainly from his studies with drosophila (fruit fly). He demonstrated the possibilities of alterations in the genetic composition of populations by changes in the relative position of genes in chromosomes, by deletion or duplication of genes, by doubling the number of chromosomes, or simply by changes in genes frequencies due to isolation of colonies with small number of individuals. The work of Dobzhansky valorized Darwin's ideas about natural selection as a decisive step in evolution. Before coming to Brazil, he had studied species from temperate-zone regions, with accentuated climatic alterations, that require organisms to adapt to different environments. The opportunity to come to Brazil allowed him to study what happens in terms of micro-evolution in a tropical place. Thus, he was interested in spending some time at Dreyfus laboratory, in São Paulo, and in collecting drosophila species in the Amazon. His coming to Brazil was a result of financial support given to his group at the University of Columbia by the Rockefeller foundation.

During his vacation and supported by his father, Freire-Maia spent a week as a trainee at Dreyfus's laboratory, under Dobzhansky's supervision and having Crodowaldo Pavan and Brito Cunha as graduated colleagues. At the quiet Alameda Glete, the laboratory was at that time a place of scientific effervescence.

After his traineeship in São Paulo, Newton returned to Minas Gerais with two genetics books and many glass bottles with drosophilae. First he went to Boa Esperança, to spend the rest of his vacation. When Newton showed the glass bottles to his father saying: "Dad, these are pure breeds of flies!" his father replied smiling: "I rather you'd brought some pure breeds of cattle [...]." He was surely joking and remembering his own father whose footsteps his son was now following.

In Alfenas, Newton had at his disposal a room equipped with a microscope and an improvised amplifying glass. He would grow flies, prepare microscope slides and show them to the students saying: "Come see the politenic chromosomes of drosophila, it's the first time it is done in Minas Gerais!"

In 1945, Newton went back to his traineeship in São Paulo for a month. This time, Dreyfus himself sponsored his trip and invited him to work there the next year, when he would graduate from dentistry. Newton was hired as a "technician", but soon started to research and go to lectures on biology at the Natural History Course. In 1948 he married Flávia, his parents' neighbor's daughter, and started to teach at a private college in order to complement the low wages as a technician at São Paulo University. Faithful to classical music, he would rock his daughter, Regina Flávia, to sleep, in 1950, adding words to Beethoven's seventh symphony's third movement.

Dobzhansky returned to São Paulo many other times: in 1948 and in 1955 to stay for a year; in 1952 to organize a collection expedition to Belém and in 1953 to take part in the jury that gave Crodowaldo Pavan the title of professor.

Newton Freire-Maia's originality and independence were manifested during Dobzhansky's second visit and would create disagreements between them. At Dreyfus's laboratory, all were concerned about studying drosophila, including Newton, but he also intended to study domestic breeds despite the criticism of Dobzhansky, who only saw reason in studying wild flies. In addition, Newton decided to broaden his line of research and started to study consanguineous marriages.

Newton obtained data on consanguineous unions from the Catholic Church of São Paulo, where the books from the different parishes of the region were preserved, containing registers dating from the eighteen century. He wanted to study the evolution of the frequency of consanguineous marriages and obtained samples that included all social classes, from rich to slaves.

In his first paper on Human Genetics, published in 1950 at *Cultus*, Newton is very critical about the eugenics movement and, using the mathematical formulations of population genetics, shows that eugenic measures were not efficacious at reducing the frequency of deleterious genes. This publication sets the beginning of human genetics as an activity of a scientific community in Brazil.

6. At the UFPR

Meanwhile, in Curitiba, at the newly founded Faculdade de Filosofia, Ciências e Letras da UFPR [College of Philosophy, Science and Letters of the Federal University of Paraná] Father Jesus Moure, a zoology professor specialized on bees and one of the rare priests to defend the theory of evolution, thought that someone from Dreyfus team should be hired with the aim to create a genetics research center there. Father Moure used to go to São Paulo and to visit the Laboratory at Alameda Glete, so he convinced the professor in charge of the Biology Department that Newton would be a good choice.

Newton began to work at UFPR in 1951, and created there the Genetics Laboratory which became afterwards the Genetics Department which offers nowadays Master and Doctoral programs. Once in Curitiba, he continued his work on the two areas of research, traveled to many Brazilian states colleting both drosophilae and data on consanguineous marriages.

Freire-Maia's first international appearance happened in 1952, when he published a paper on consanguineous marriage in the *Ameri-*

can Journal of Human Genetics, following the publication of a short note, the year before, in the Brazilian journal *Ciência e Cultura*.

Newton was a pioneer regard Human Genetics, Medical Genetics and also Genetic Counseling in Brazil.

The Rockefeller Foundation offered generous support to Freire-Maia's laboratory as well as the laboratories in São Paulo and Porto Alegre. It also paid scholarships to several young geneticists who turned to human genetics and supported the publication of the Boletim da Sociedade Brasileira de Genética [Bulletin of the Brazilian Genetics Society] as well as scientific meetings.

According to Father Moure, Freire-Maia used to work seven days a week even when his formal contract was not full time. He demanded the same dedication from all who came to work with him and was very strict about discipline and use of public money.

At the end of the 1950s, Freire-Maia enrolled the doctoral program on Natural Sciences at the Faculdade de Filosofia, Ciências e Letras da Universidade do Rio de Janeiro [College of Philosophy, Science and Letters of Rio de Janeiro University]. Despite financial difficulties, he traveled several times from Curitiba to Rio for exams and eventually to defend the thesis on *Consanguineous Marriages in Brazil*, with data collected in the south of Minas Gerais State. The doctoral degree was obtained in December, 1960.

7. Conversion

When I started to work with him, during the Master's program, I noticed among his students and colleagues mixed feelings of admiration and fear, although no one questioned his leadership or doubted that the department owed its status to him. He used to, for instance, catch someone distracted in the corridor and ask, all of a sudden, some genetics concept and, if the victim did not answer immediately he shouted something of the kind: "If you do not know genetics you can not belong to this department!" He was like that, straight forward and abrupt, but if contested, he would discuss as equal and when proved wrong, apologize sincerely. He was innocently vain but not arrogant. As time went by, he became more and more gentle and, because this change in behavior became more noticeable after his wife's death and his second marriage, with the geneticist Eleidi Chautard, his friends used to say that she had "tamed the beast". One evidence of his strict but democratic and father like leadership is that his team workers soon created their own independent lines of research, forming teams of their own, with his support. With time, the feelings of old members of the department ripened and became purely of admiration and love towards the old master.

When I met Professor Newton in the 1970's he considered himself as an atheist and I regarded his disposition to participate in debates about religion and science, and his invitations to rabines and Catholics for the department seminars as a demonstration of respect for other peoples' ideas. I heard him saying once something like that: "People have their right to think whatever nonsense they want…" Nevertheless, the subject is not simple, for he himself would latter state that at that time he was "an atheist looking for God." I recall him in one of our daily chats during the department coffee brake saying to a Marxist student: "It doesn't matter that you don't believe in God, cause He believes in you."

His conversion to Catholicism, in the early 1980's, received different interpretations by his friends, but if one considers his own story, told during interviews and publications, the conclusion is that he became a believer when he managed to conciliate his ideas about science and one religion.

Some extracts of his book "Creation and Evolution – God, chance and necessity", show that his faith emerged out of the possibility of bridges between science and religion.

> I have spent more than half of my life under the sign of atheism or agnosticism, which I considered **inevitable fruits of my position as a scientist**. When I was 20 years old, I already considered myself an atheist – a militant one, very keen on breaking other people's faith… For me, only an obtuse stupidity or a profound ignorance could coexist with faith. Cultivated faithful men – and I knew that their number was growing – could only be like that as a result of some perturbation of world-view.
>
> In 1954, I was profoundly touched by Thomas Merton's book *The Seven Storey Mountain* and spent months praying so that God (in whom I did not believe but, at the time, who I ardently wanted to exist) gave me faith. In vain: faith did not come.
>
> […] So, my conversion lasted, 26 long years. The process got faster during the tree years preceding 1980. I have read, thought, and heard a great deal. And finally, on March 25th, 1980, I confessed with Friar

> Benjamin Beticelli and received communion for the first time in 45 years.
>
> It remains, among scientists; the belief more or less general, according to with science is not compatible with faith [...]. Nevertheless, time has shown that **the conflict was but apparent and so, doubts were resolved and points of disagreement settled** (Freire-Maia, 1986, p. 96).

In an interview published by *Ciência Hoje*, in 1988, he explains how he accommodated his religion with his vision of science:

> Religion has accumulated a series of superstitions along the centuries, therefore the conflict between [religion and science]. A feigned conflict. It is not the role of religion to give explanations about the origins of universe, living beings or mankind. This is a scientific problem [...]. **My viewpoint change did not alter my research projects**. I believe today in the same scientific theories I accepted before [...]. The Vatican cannot give opinions about the origin of the cosmos [...]. It requires courage to state that religious tradition is wrong in these issues! It is impregnated with superstitions and must be analyzed from the viewpoint of mythology.

8. Political views

During the period 1940-1945 Newton was sympathetic with communist ideas but latter on he became interested in the democratic socialism and even run – and was defeated – for a place in the parliament. In 1964, when the *coup d'état* deposed President João Goulart, Newton was giving a conference about the University at the Casa do Estudante Universitário [House of the University Student]. With the military regime, interrogations, dismissals, rights abrogations, tortures and murders began. Many intellectuals, geneticists among them, were chased. As in other universities, the UFPR established a commission with the aim to investigate supposedly "subversive activities" by professors, students and staff. One morning, Newton was surprised with a short notice in the local newspaper announcing that the commission had decided to dismiss him from his job. When he was called to speak before the committee he declared: "If it is a crime to be a socialist, then I am a criminal!" And he went home frightened and surprised with his boldness, sure that dismissal was on the way. Nothing happened but anxiety remained with his family for some time. "And we carried on the usual life, short of money, with fear and hope." In the same year, Newton was invited to speak about methods used to estimate genetic effects of consanguineous marriages during a symposium at Cold Spring Harbor, but the Dean did not grant him permission to leave the country. His friends and colleagues tried to persuade the Dean to change his decision without success. Newton had to return the tickets sent from United States and could only mail his work for publication in the event annals. He recalls: "It was the beginning of the national shame at the international level [...]."

In 1968, after the famous congress at Ibiúna, the UNE [National Union of Students], became extinct by a government edict but the students used to gather in secrecy. In Curitiba, the president of the local union was arrested with some other students, as a consequence of these illegal activities. A group of students then went to professor Newton's home and asked him to use his influence and plead for the students' liberation. Although fearing being arrested for defending people considered subversive, Newton agreed. The students, concerned about his security, ended up finding another person, but remained thankful to him.

In 1980, his name would be used in the struggle for the UFPR democratization. After retiring, he kept working as a volunteer and, five years after, was engaged as visiting professor. Newton was a symbol at the graduate program and the department wanted to keep his name on the list of professors. When his contract expired, his colleagues tried to renew it but it was denied. Everybody knew that the denial was legal, but they also knew that exceptions were made for other people and that the decision was political, since most department members were left wing and opposed to the university administration. Freire-Maia said at the time: "I do not need this, I have a grant from CNPq", to which his colleagues replied: "But we need you as a flag, as a reason to fight." The movement received support from scientists of other departments and of several other places in Brazil, from innumerous scientific, cultural and community societies, from the mayor, students, politicians and journalists. The conflict between the high administration and the opposition went on for several months, during which nine caricatures and more than 120 articles were published in newspapers and magazines about the "Freire-Maia affair".

Meanwhile, the direction of the professor's union, aligned to the central administration, was put in question, culminating with the election of another board of directors. Newton's contract was not renewed but the authoritarian group that directed the university lost support. The UFPR, since then very conservative, begins a new era. Deans start being now elected by the university's community, as well as the heads of departments. The union becomes one of the most active in the country.

This period was tense and energy consuming to everybody. Accusations were made, to the "insubordinate" and their families and the peace of mind showed by Newton all the time was taken, by some of his friends, as a result of his newly acquired religiosity.

Newton Freire-Maia used to say: "I was born in 1918, a great year, a century after Karl Marx and half a century before AI-5." He meant 5th Institutional Act, which withdraw political rights from opponents to the dictatorship installed in 1964. He died in 2003, time also of Stephen Jay Gould's and Ilya Prigogine's deaths, In his eighties, he was still very active, influencing people and completing his last book, on the truth of science and other truths.

With the aim to situate Newton Freire-Maia in the history of Brazilian science, I will comment on the scientists who, together with him, form the first generation of Brazilian Human Geneticists. This account will be preceded by a few notes on researches related to human genetics in the country, by individual researchers who did not constitute a community of geneticists for they did not have genetics as their area of main interest.

9. Brazilian human genetics previous to Freire-Maia

Research and teaching activities related to genetics first happened in Brazil in the period 1920-1950. The subject was studied in the agricultural schools and medicine schools, within the pathology discipline.

During the 1920s, studies related to human heredity were done by the eugenists. Renato Kell, a physician and pharmacist, was the leader of eugenics in Brazil that, unlike the Anglo-Saxon tradition, considered the environment among the causes of race improvement because it had Lamarckian postulations (Castañeda 1998). This strong influence of Lamarckism among Brazilian intellectuals and the position of the Church against birth control are some of the peculiarities of the Brazilian eugenics (Salzano 1992). The interdiction of interracial marriages and sterilization of handicapped was emphasized in Europe while here the main concern was the control and prevention of diseases considered hereditary at the time, such as alcoholism and syphilis. Nevertheless, some papers with a Nazi approach were published during the 1930s in the Eugenics Bulletin edited by the Brazilian Eugenics Commission (Beiguelman 1990).

The movement climax happened in 1929 with the First Brazilian Eugenics Congress, in Rio de Janeiro, with Roquete-Pinto as the chairman and the participation of 200 people: general practice doctors, legist doctors, psychiatric doctors, journalists and parliament representatives, besides delegates from Argentine, Peru, Chile and Paraguay (Stepan 1990).

The consequences of Nazi measures during World War II were the discredit of eugenics. Studies on human heredity would only become important again in the 1950s, under the Mendelian and Darwinist postulations. Yet, several Brazilian workers studied the heredity of human diseases and the frequency of bloods groups during the 1920-1950 period (Beiguelman 1979, Azevêdo 1988). Some of these scientists were from Bahia State. The first Brazilian work on blood groups to be published was by Octavio Torres, in the early 1930s, at *Gazeta Médica da Bahia*, a journal that circulated between 1866 and 1976. Jessé Accioly described, in 1947, at a local journal of Universidade da Bahia [Bahia's Federal University], called *Tertúlias Acadêmicas*, the mechanism of hereditary transmission of falciform anemia. He was the first to find that Mendelian laws applied to this disease, but it was James Neel's work, published in the United States with the same conclusion, that became notorious (Beiguelman 1981). Later on, in 1973, a study by Eliane Azevêdo at the *American Journal of Human Genetics* acknowledged the fact and Jessé Accioly got international recognition for his pioneer work. These scientists' production was important but their works on genetics were sporadic and published at journals of very low impact, so they did not constitute a community of geneticists. The few teaching and research centers working with genetics in Brazil at the time were geared to plant genetic improvement.

10. Brazilian human genetics – generation one

The first agreements between the Rockefeller Foundation and the Medical School of São Paulo were settled in 1918 (Marinho 2001). They enabled the substructure conditions and full-time jobs that made it possible for professionals to give rise to a strong line of research on drosophila population genetics at the 1940s. This group disseminated to other regions and other lines of research throughout the country, human genetics between them. These agreements involved, on behalf of the Foundation, the construction of new buildings for the medical school – which would latter become part of the Universidade de São Paulo [São Paulo State University], besides scholarships and support for installation and maintenance of laboratories. The Brazilian counterpart involved hiring full time lecturers, limiting the number of students according to laboratory and clinic conditions as well as concentrating all sectors of the school in one campus.

The strong relations between the Rockefeller Foundation and the Brazilian geneticists were due, among other factors, to the efforts of Harry Muller Jr., one of the Foundation directors. He played an intermediary role in the various trips of Dobzhansky to Brazil, during the period 1943-1953 and also in the financial resources that made it possible to consolidate the human genetics laboratories. Muller's retirement, in 1961, coincided with this support decline (Salzano 1992). At the end of the decade, when the Foundation support ceased, the CNPq [National Research Council], created in 1951, was consolidated and in 1971 was created the FNDCT [National Fund for Scientific and Technological Development], and soon later the FINEP [Studies and Projects Financier].

André Dreyfus, the head of the Biology Unit at Universidade de São Paulo, had an important role in the events described above. It was in his laboratory that Dobzhansky worked and left followers. As early as 1927 he already included mendelian genetics and Darwinian evolution in his histology and embryology classes at the Medical School.

All the followers of Dreyfus and Dobzhansky began their research careers by studying drosophila, but several of them migrated to other areas. Among the ones that turned to human genetics is Newton Freire-Maia, soon followed by Oswaldo Frota Pessoa and Francisco Mauro Salzano. Pedro Henrique Saldanha and Cora de Moura Pedreira, who also belong to the first generation of Brazilian human geneticists, were not followers of Dobzhansky.

Oswaldo Frota-Pessoa told me that he began to work with human genetics under Newton's influence, as soon as he got in touch with Gunnar Dahlberg's work on Genetic Isolates and became a contagious enthusiast. When he was 21 years old, Frota had an essay entitled "Why children resemble their parents" published in a popular magazine. This was the first demonstration of his talent to write about science to the general public. He is the author of hundreds of essays in newspapers and popular magazines as well as many textbooks. He was first a teacher in Rio de Janeiro, then a lecturer at Universidade do Brasil and finally at Universidade de São Paulo. Frota has given an important contribution to the Latin American science education by means of his innumerous textbooks about science, biology and education. He has created the Laboratory of Human Genetics at the Biology Unit of São Paulo University, directed research projects in different areas of human genetics, supervised many theses and, just before retiring, began working with psychiatric genetics.

Pedro Henrique Saldanha started working with human genetics when he was a Natural Sciences student and had Frota as his professor. He corresponded with Gunnar Dahlberg and would later choose the subject of his doctoral thesis after a suggestion by the Swedish geneticist. He published his first paper on human genetics in 1954, when working as a teacher in a small town of São Paulo State. Saldanha has introduced the first regular course on human genetics in Latin America at the Medical School of Universidade de São Paulo. There he created the Laboratory of Medical Genetics. Afterwards, he stimulated the organization of various courses, laboratories, disciplines and departments of human genetics in several colleges and medical schools. He did not work with Dobzhansky but received funds from the Rockefeller Foundation from 1957 to 1967, a scholarship during the doctoral studies at Universidade de São Paulo (1957-1959) and a grant to work as visiting professor at the medical school in Michigan (1960-61).

Francisco Mauro Salzano met Freire-Maia in 1951, when he was newly graduate at Porto Alegre and went to São Paulo in order to follow doctoral studies. At that time Freire-Maia was leaving for Curitiba and Salzano occupied his desk in the laboratory. The two of them used to chat during Newton's visits to São Paulo and became closer when Salzano went to Ann Arbor with a Rockefeller grant and Newton was already there. The friendship and professional collaboration resulted in papers and books published together. Salzano has worked and supervised works in very different areas of genetics but his main works are on human genetics, mainly with Indian populations. As well as Freire-Maia at UFPR and Frota-Pessoa at Universidade de São Paulo, Salzano became an admired leader at Universidade Federal do Rio Grande do Sul [Federal University of Rio Grande do Sul State], with a strong influence on the education of human geneticists nowadays spread in research centers in Brazil and abroad. Like the other pioneers, Salzano has received important prizes and is often mentioned as an example of love for science.

The first generation of Brazilian human geneticists includes Cora de Moura Pedreira who, together with Eliane Azevêdo and Lucy Peixoto established human and medical genetics in Salvador, Bahia, at the same time as the Dobzhansky group originated the groups at São Paulo, Paraná and Rio Grande do Sul.

Cora graduated from Medical School in 1938 and defended her doctoral thesis in 1954 working with blood groups in the population of Salvador. She has studied Indian communities in several states in the north and northeast of Brazil and standardized human and animal cytogenetics. She has created the Cytogenetics Laboratory of the Biology Unit at the Universidade Federal da Bahia [Federal University of Bahia] and worked intensively with chromosomal studies of newworld primates. Her laboratory still offers genetic advice to the population of Bahia and, thanks to her enthusiasm, other laboratories were created, such as the Plant Cytogenetics one.

Her teaching activities were also intense and diverse. In 1967 she coordinated a genetics short-term course supported by CAPES and the Ford Foundation and was also in charge of the introduction of the discipline of medical genetics in the Medical School in Salvador, in 1968.

Cora developed a pioneer work in many aspects and influenced several generations of students, lecturers and scientists at the Federal University of Bahia, as well as students and lecturers of different universities in the country and abroad.

Unlike what happened in São Paulo, Curitiba and Porto Alegre, the group in Salvador did not originate from drosophila works and is not mentioned in the essays already published about the history of Brazilian human genetics. Nevertheless, Cora knew Dreyfus, visited his laboratory during one of Dobzhansky's visits, used drosophilae in her classes and participated in the creation of the Sociedade Brasileira de Genética [Brazilian Genetics Society] in 1955.

The community of human geneticists in the country, in a number that exceeds 300 researchers among the total of approximately 2,000 geneticists, had its origins in this group of five persons.

11. Final comments

This work focused on a person that had an important role in the establishment of Brazilian human genetics, a science that has reached a leading position in Latin America and kept its identity as scientific production with national and international recognition.

The support of the Rockefeller Foundation, followed by that of CNPQ [National Council for the Scientific and Technological Development]; of SBPC [Brazilian Society for the Advancement of Science] and of SBG [Brazilian Genetics Society] were essential in the development of Brazilian human genetics. But the passion scientists felt for their work and their refusal in accepting to do a periphery kind of science were equally important.

Most Brazilian scientists in mid-twentieth century had a feeling that Science could contribute to the population welfare. Newton Freire-Maia was one of them. Discussions about the complexity of the relations between scientific development and social and economic development were not so present in his discourse. Nevertheless, the way he behaved as a person and as a professional shows that he was concerned about ethical and social issues.

The importance of Newton Freire-Maia goes beyond genetics, in such a way that a research done by FAPESP in 2000 places him among the 150 Brazilians who helped building science and technology in the country.

Studies such as the present one, dealing with very recent times, have as a restriction the lack of withdrawal in terms of time and emotions, which would allow for an epistemological analysis that should substantiate a text on the history of science. Such studies, however, owe their relevance to the fact that they serve as basis for further works. During this kind of study important documents are produced or obtained which otherwise might be only regarded as old paper and be lost in personal or institutional archives. Besides, a research that focuses on a period near to the present and on live characters makes it possible to obtain oral testimonies, a valuable tool to preserve points of view of the people who are in the process of making history.

References

Azevêdo, E.S. (1973), "Historical Note on Inheritance of Sickle Cell Anemia", *American Journal of Human Genetics* 25: 457-458.

Azevêdo, E.S. (1988), "Genética Humana no Brasil: Passado e Presente", *Ciência e Cultura* 41 (5): 439-466.

Beiguelman, B. (1981), "A Genética Humana no Brasil", in Ferri M.G.E & S. Motoyama (eds.), *História das Ciências no Brasil*, São Paulo: EDUSP, pp. 273-306.

Beiguelman, B. (1990), "Genética e ética", *Ciência e Cultura* 42: 61-69.

Castañeda, L.A. (1998), "Apontamentos historiográficos sobre a fundamentação teórica da eugenia", *Episteme* 3 (5): 23-48.

Ferrari, I. (1988), *Homenagem às pioneiras da genética na Bahia*, Feira de Santana: Livreto publicado pela editora da UFFS.

Ferrari, N. (1997), "Breve História da Fundação Rockefeller e de seu Papel no Desenvolvimento da Genética Humana Brasileira", in *Anais do VI Seminário Nacional de História da Ciência e da Tecnologia*, Rio de Janeiro: SBHC, pp. 479-484.

Ferrari, N. (2000), "O Departamento de Genética da Universidade Federal do Paraná na Origem da Genética Brasileira", in *Anais do*

VII Seminário Nacional de História da Ciência e da Tecnologia, São Paulo: SBHC/EDUSP, pp. 169-173.

Freire-Maia, N. (1950), "Eugenia e genética de Populações", *Cultus* I: 1-9.

Freire-Maia, N. (1951), "Casamentos consangüíneos em populações brasileiras", *Ciência e Cultura* 3 (4): 283-284.

Freire-Maia, N. (1952), "Frequencies of consanguineous Marriages in Brazilian Populations", *American Journal of Human Genetics* 4 (3): 194-203.

Freire-Maia, N. & I.J. Cavalli (1978), "Genetic Investigations in a Northern Brazilian Island. I- Population Structure", *Human Heredity* 28: 386-396.

Freire-Maia, N., Andrade, F.L., Athayde-Neto, A., Cavalli, I.J, Oliveira, J.C., Marçallo, F.A. & A. Coelho (1978), "Genetic Investigations in a Northern Brazilian Island. II- Genetic Drift", *Human Heredity* 28: 386-396.

Freire-Maia, N. (1986), *Criação e Evolução - Deus, o acaso e a necessidade*, Petrópolis: Vozes.

Freire-Maia, N. (1995), *O que Passou e Permanece*, Curitiba: Editora da UFPR.

Frota-Pessoa, O. (w/d), *A Rambling Rationalist* (manuscript autobiography).

Frota-Pessoa, O. *et al.* (1988), "O Acaso na vida do pesquisador", *Ciência Hoje* 9 (49): 16-22.

Glick, T.F. (1944), "The Rockefeller Foundation and the Emergency of Genetics in Brazil", in Cueto, M. (ed.), *Missionaries of Science-The Rockefeller Foundation and Latin America*. Bloomington: Indiana University Press, 1944, pp. 149-164.

Revista da FAPESP - Fundação de Amparo à Pesquisa do Estado de São Paulo, n° 52, abril de 2000.

Marinho, M.G.S.M.C. (2001), *Norte-Americanos no Brasil. Uma história da Fundação Rockefeller na Universidade de São Paulo*, São Paulo: Editora FAPESP.

Saldanha, P.H. (1954), "Taste Tresholds for Phenilthyoureia among Students in Rio de Janeiro", *Revista Brasileira de Biologia* 14 (3): 285-290.

Salzano, F.M. & N. Freire-Maia (1967), *Populações Brasileiras, aspectos demográficos, genéticos e antropológicos*, São Paulo: EDUSP.

Salzano, F.M. (1992), "The history and development of Human Genetics in Brazil", in Doonamraju K.R. (ed.), *The history and development of Human Genetics-Progress in different countries*, Singapore: World Scientific, pp. 228-255.

Stepan, N.L. (1990), "Eugenics in Brazil 1917-1940", in Adams, M.B. (ed.), *The Wellborn Science. Eugenics in Germany, France, Brazil and Russia*, Oxford: Oxford University Press, 1990, pp. 110-152.

Interviews with: Antonio Brito da Cunha, Crodowaldo Pavan, Eliane Elisa de Souza Azevêdo, Newton Freire-Maia, Eleidi C. Freire-Maia, Euclides Fontoura Jr., Iglenir J. Cavalli, Remy Lessnau e Riad Salamuni.

The Beginning of Tropical Medicine in Argentina and Brazil

Sandra Caponi

Department of Public Health, Federal University of Santa Catarina (UFSC)

In this paper, we aim to analyze the way by which the early bacteriological investigations developed in Brazil and Argentina were integrated with a new kind of medicine born in the 1890s and concerned with the role played by vectors in the transmission of tropical diseases. A specific kind of disease is then considered, one of which the model is paludism or malaria, and the study of new transmission channels is initiated: the living intermediate agents, fundamentally the arthropodes (articulated invertebrates), capable of interfering in many ways on disease propagation. There is not necessarily a clearly derived relationship between bacteriology breakthroughs and the appearance of arthropods as vectors. Similarly, there is no absolute continuity between those prophylactic measures imagined and proposed by post-Pasteur hygienists (sanitation, disinfection, vaccines) and the specific prophylaxis required to fight every arthropod recognized to be a vector.

As stated by Canguilhem:

> In fact, it seems very simple to make the distinction, in an epidemic disease, between focus, specific agents, form of transmission and diffusion [...]. However, the concepts of germ, vehicle, and intermedi-

ate host asked for a laborious research involving observation, analogies, experimentation and refutations (Canguilhem 1989, p. 14).

Among those vector-transmitted diseases we find, for instance, the viral diseases caused by *arbovirus* (arthropod-born virus), as is the case of yellow fever; bacterial diseases, such as paludism and trypanosomiasis involving protozoary parasites, in the first case the Plasmodium and in the second the trypanosoma Cruzi; finally, diseases such as filariasis which involve the parasitic helmints.

It is my intention to analyze conceptual and institutional conditions for the emergence of Latin-American tropical medicine. A draft will be presented of an epistemological historical analysis, not only about the way new concepts and theories have been developed, but also about the different historical moments when two research traditions faced each other. A retroactive look, both internal and external, of the history of sciences, can offer a better understanding of the arguments presented by those who developed the theories.

Two possible accounts exist for this study: the first, which is a more classic one, speaks of continuity and perfecting of research programs and studies performed by bacteriologists and microbiologists. By the end of the XIX century and beginning of the XX century, the most prominent bacteriological institutes, Koch and Pasteur, sent investigators to the African and Asiatic colonies to survey the kind of study performed on metropolitan laboratories for tropical areas. Faced with the threat of new diseases affecting the white population who intended to live in the tropics, it can be said those protocols were improved, so as to answer to the new challenges (Stepan 1976, Michel 1991, Löwy 1991, Darmon 1999).

The other account speaks of a new study universe, having research on malaria as a model. Here a new theoretical field and a new scientific discipline emerge: the tropical medicine, to be born when Patrik Manson founded "The London School of Tropical Medicine", in the year of 1898. In the latter case it is said a confluence existed between the new microbial studies, the classic studies on warm climate diseases, and the studies on entomology and parasitology. Research protocols, upon which the metropolitan theory of germs was based, proved insufficient, and had to be modified (Arnold 1996, Peard 1996, Power 1998).

It is our interest to ask how these theoretical and practical questions influenced the development of Latin-American medicine in the beginning of the XX century and final part of the XIX century. Of particular interest is to analyze the manner how Argentinean and Brazilian researchers built their research programs. To which diseases did they give priority? How did they understand, and how did they accept the studies aimed to those tropical diseases which, at that time, decreased Argentinean and Brazilian populations? How did they intend to face recurrent epidemics of yellow fever and malaria which were so much concerning European researchers? Which research protocols were privileged? A punctual analysis of some of the arguments used by Brazilian and Argentinean hygienists, sanitarians and researchers between the years of 1890 and 1916, regarding explanatory models and prophylaxis strategies to be adopted face to diseases such as yellow-fever, malaria and *trypanosomiasis americana*, known after 1909 as Chagas' disease, will make possible to observe two coexisting research models which, in one same historical moment, coexist and are rivals.

The analysis of this moment when two explanatory strategies are in confrontation shall make possible to understand in which sense we are able to talk about two models for the intelligibility of the diseases, or to differentiate "visibility spaces": on the one hand, the study of tropical medicine and vector-transmitted disease; on the other hand, the extension to the tropics of a classical manner to understand diseases and the classical strategies of prevention for which hygienists fought, associated to the new results in microbiology.

In Brazil, both explanatory models have been adopted by Brazilian researchers as complementary strategies. As for Argentina, it will give priority, at all moments, even for tropical diseases (malaria, yellow fever, and trypanosomiasis), to the research program started by Pasteur and by Koch.

This theoretical and epistemological difference generated debates and scientific controversies between the two countries at the moment international measures of prevention had to be defined such as demonstrated on the Annals of the First Latin-American Medical Congresses of 1904, 1907 and 1913. During the years 1890, Argentina developed its plan for urban reorganization, housing control, and popular diseases such as syphilis and tuberculosis. In the following years, the concerns of Argentinean researchers would be centered on the building of laboratories and an Institute of Bacteriology.

These bacteriologic studies assure the scientific legitimacy of the work developed by the classic hygienists Guillermo Rawson (1891) and Eduardo Wilde (1885). The new hygienists of the first decades of the XX century, among which are mainly the names of Emilio Coni (1918) and José Penna (1904), will enhance statistical studies, centering all their confidence in microbiology conquests. Statistical studies joined microbiological researches to perfection. The idea was to isolate and discover new microbes, to develop specific vaccines and sera, and give continuity to the classic steps of disinfection, sanitation and urban reorganization. Buenos Aires was, at that time, a model of hygienic city to be copied by other Latin American capitals.

The main concern of these hygienists was centered on diseases stemming from urban over-crowding and life conditions of popular classes: tuberculosis, syphilis, and alcoholism. However, Argentina was not limited to a sanitized Buenos Aires: in those days, same as in our days, there were extremely poor cities in the rest of the country, permanently threatened by diseases typical of warm climates.

Nevertheless, as a lingering consequence of the two dramatic epidemics of 1871 and 1890 in Argentina, vector-transmitted diseases kept being studied using the same strategies as any other directly transmitted infectious disease: sanitation, disinfection, and the production of vaccines and sera.

After 1903, Brazil had the largest center for bacteriological studies in Latin America, that which is now the "Oswaldo Cruz Institute". In it, in addition to producing sera and vaccines, research and prophylaxis programs were developed for the great epidemics which concerned Brazil: yellow fever, malaria, plague, and smallpox. Three of these are vector-transmitted, requiring differentiated research protocols: medical entomology and zoology studies. Also required are specific prophylactic strategies for vector fighting and control: rats, fleas and mosquitoes. Another epidemic, of urban-rural character, is added to the list after the year 1909: *trypanosomiasis americana* or Chagas' disease, also included among vector-transmitted diseases, in this case by the *barbeiro* or *vinchuca*.

Although the epidemics threatening Argentinean and Brazilian populations were more or less identical (not only smallpox and tuberculosis but also yellow fever, plague, paludism, and Chagas'), the research and control strategies developed by each of these countries were completely different. In Argentina, greater emphasis was given to the production of vaccines, in conjunction with measures of cleaning and sanitation; in Brazil these measures had to be supplemented with other studies, not exclusively laboratory ones.

Identification of the possible vectors required studies on natural history and entomology to make possible the classification, systematization and localization of arthropods.

To be able to understand the existing differences between these two theoretical approaches, in countries having similar characteristics, it might be illustrative to understand how the idea of "tropic" was built and modified and, fundamentally, how the Pasteur Institute organizes its research overseas or, still, which are the research protocols with which it starts work at the French colonies in Africa and Asia.

Everything seems to indicate that the notion of "tropic" was vested with a symbolic meaning, rather than a physical one. It defines something that, for Europeans, appears as their "other", something culturally, topographically and politically different from Europe. Facing this threatening alterity, the temperate regions acknowledged their positivity. In developing this notion, a contribution was received from certain theoretical certainties strengthened in this period. Firstly, the statistical studies demonstrating, based on quantitative data, the extreme vulnerability of the white population as compared to the local ones, reflected on the differentiated rates of mortality. Secondly, the development of geography and medical topography gave legitimacy to the idea of the existing local causes, linked to a specific topography, vegetation, insects and animals capable of producing or transmitting certain diseases.

Such a hypothesis made possible, and required, the study and the classification of a wide variety of fauna, flora, soil and topography of extreme scientific precision, allowing for an intuitive correlation with local pathologies. Finally, it must be emphasized on this program the persistence of miasmal theories and the recovery of Hippocratic writings speaking of the particular danger posed by hot air.

At that time, the relation between climate and geography, under the influence of thinkers like Montesquieu, seemed to keep a direct link with the characteristics of the different human societies. Climate would define the kind of man and society, their morality and their political capacities. An absolute continuity was seen to exist between climate, morality, and pathologies. Consequently, doctors and hygienists from the 18th century and first half of the 19th understood that, to be able to imagine a medical and moral transformation, it would be indispensable to establish a clear connection between conducts and the physical medium.

To be able to understand this continuity, it becomes necessary to speak of a general "epistemological ground" on which it is not yet possible to speak of a differentiated social space and of a natural space; little by little, along the XIX century, they are thought of as autonomic spaces, requiring from each one a definition of its own rules and objects. In the specific case of Brazil, whenever climatic medical explanations were accepted, the country was lead to a sort of historical pessimism: "the tropical climate was made responsible for endemic and epidemic diseases in the country. It was also assumed the Brazilian population, racially crossbred, was sensuous and passive, susceptible to diseases and incapable of control and individual or collective rationality turned to progress and civilization" (Stepan 1976, p. 63).

These ideas that the tropics condemned Brazil to disease and backwardness were multiplied among doctors and hygienists of the XIX century, as seen on this text from 1850:

> Those who inhabit marshy countries are weak, their skin is colorless or yellowish, the flesh is soft and without plasticity (sic.*elansterio*), imbibed with serum. They exhibit a repulsive swelling; their eyes lack expressiveness (...), they have reduced height and are of flawed conformation (...); the influence of "paludical emanations" upon the moral makes a man a libertine. Among women, a greater rate of abortions and infanticides can be seen (Ferreira França 1850, p. 1, *apud* Machado 1978).

The image of the first travelers, who praised the exuberance and beauty of the tropics, was changed at the end of the XVIII century: "a kind of negative representation of the tropics, ultimately exotic, came to be common on medical literature, particularly regarding Western

Africa and Western India" (Arnold 1996, p. 7). The malevolence of the tropics became a medical theme, from frightening storms and voracious animals, by extension or analogy, one arrived to the extreme seriousness of the diseases concentrated in those regions. In the concrete case of Brazil,

> by the end of the XIX century the picture of Brazil as a tropical paradise had already disappeared, and the climate was on the head of most people as the main cause of diseases and also as the main obstacle to the enhancement of civilization in the country (Stepan 1976, p. 54).

It is true that, during a large part of the XIX century, Brazilian hygiene seemed to reproduce the discourse of the large cities. As an unquestionable truth it was accepted that the hot climate determined limits to the advancement of science and culture. What Nancy Stepan called racial pessimism seemed "to confirm the belief of many European anthropologists that Brazilian racially mixed population and/or tropical climate condemned them to disease and backwardness" (Stepan 1976, p. 26). This thesis remained in the mind of doctors and local intellectuals during a substantive part of the XIX century.

A positivist doctor (Pereira Barreto 1890) started then to support the need to foster scientific studies in Brazil, particularly those covering control and contention of epidemics such as the yellow fever. Two things contributed to that transformation:

Firstly, the installation of the *São Paulo and Rio de Janeiro Institute of Bacteriology,* directed by Adolfo Lutz and Oswaldo Cruz. Work on this matter made possible to show that many of the diseases attributed to the torrid climate had specific causal agents, and could be classified in the international listing of already known diseases: cholera, tuberculosis, and typhoid fever.

Secondly, entomological studies of classification and the identification of local arthropods were started, initially by Oswaldo Cruz and then by Adolfo Lutz and Carlos Chagas, among others. At this moment, it was no longer of great interest to show that in Brazil the same diseases could be seen as in Europe, nor to say that the same bacteriological agents existed, but rather to observe the peculiar character of certain diseases requiring the intermediation of vectors with very specific characteristics, as they inhabited (or mostly inhabited) the tropics. Such unknown species required a careful examination of their anatomo-physiological character, of the spaces they inhabited, their habits, etc...

It seems that classic accounts that speak of the transposition of research programs of European microbiology to Brazil were met with some difficulties. This classic story of science maintains that this net of relationships between climate, local geographical and physical particularities, and pathologies which are peculiar to hot climates, comes to disappear at the end of the XIX century. These historians will then argue that the climatic explanations lose their importance and are replaced by modern explanations founded on microbiology. Thus, Nancy Stepan states that, after Oswaldo Cruz, miasmal and climatic explanations are to be abandoned in the search for specific causal agents: germs and bacteria. Contrary to that, analyzed documents seem to show that in spite of the coincidence between new ideas in the commencement of the century showing Brazil as the future world power and the interest on microbiology, it would be wrong to imagine only the introduction of bacteriological studies in Brazil by Oswaldo Cruz or those who preceded him (Benchimol 1999) could have been the cause of control of "mild climate" diseases. Let's not forget that, in the case of yellow fever, the specific microbiological agent was kept a mystery until 1930.

Without the knowledge of entomological particulars of the *Aedes-aegypti* (then known as *Culex*), microbiological studies could hardly have contributed to the control or reduction of this disease.

The idea of a simple transposition of research programs from central countries does not seem to be enough. In fact, when we consider the importance of studies made by naturalists in determining that the yellow fever vector is part of a given species of mosquito and not another, or that Chagas' disease is transmitted by a blood-sucking mosquito with domestic or para-domestic habits, we come to the conclusion that, more than speaking of transference or imposition we should speak of the building up of a knowledge which is the result of "synergetic relations between center and periphery" (Stepan 1976).

To be able to understand how far we can speak of an autonomous building of knowledge, it might be illustrative to analyze how the Argentinean researchers saw tropical diseases, how these scholars established links with European research programs, particularly the Pasteur Institute, and how was the articulation between studies by helminthologists, parasitologists, and zoologists (concerned with the local specificity of insects and animals) and diseases such as yellow fever, malaria, or Chagas'.

The variety of climates in Argentina, and the fact that the capital is in a moderate climate, dispelled the theses of "climatic pessimism". Fears of the tropics had then a privileged location, the across-the-border Brazil, and the only problem apparently posed by tropical diseases was the geographical closeness, favoring pestilent contagion. This account of Argentinean doctors apparently forgot that a large part of the country is in a region of sub-tropical climate, and that part of the north (the provinces of Salta, Formosa, and Jujuy) have tropical climate. In this way, the new Argentinean hygiene (an heir from microbiology) did not have to break with climate myths, nor to overcome the associated ideologies of sanitary pessimism. Microbiology sought to help, rather than negate, the interventions of classic hygienists. With Guillermo Rawson, who fought for statistics, sanitation and the mephitic exhalations, and Emilio Coni, who argued for statistics, sanitation and microbes, the continuity was complete. Little by little are to be integrated, to the still miasmatic speech of the 80s, "the eminent discoveries of Pasteur" (Rawson 1891, p. 203). Argentina will become "Pasteurized" without conflicts, being then able to present its capital as a model of sanitation, hygiene and modernity.

However, from the 1871 and 1890 epidemics until the beginning of the XX century, the most feared disease, responsible for taking the largest number of Argentinean lives, was the yellow fever. This disease has always been associated with sanitation deficiencies and also the proximity to Brazil: "the sick persons who started the hellish prowl came from Brazil" (Bellora 1972, p. 32). An argument was offered that the first epidemics of 1971 had found a poorly sanitized city of Buenos Aires, and that the new measures applied by the sanitarians helped in cutting down the seriousness of the following epidemics. This argument will be supported in 1884 by Rawson, and repeated from 1904 to 1916 by José Penna. In the beginning, there was lack of knowledge, and then open opposition to the theses of vector transmission. At no time was the necessity considered of the three main conditions for disease propagation: the diseased person; the specific agent (the virus); and the vector (the Aedes aegypti).

To be able to understand the reasons why Argentine does not take into account that possibility, it would be necessary to consider the resistance by Pasteurian researchers themselves to accept the novelty of studying tropical diseases. Argentina reproduced faithfully the Pasteurian programs, respecting its research protocols. Argentinean researchers seemingly believed their sanitary problems had no connection with the pestilent tropics, and were rather identical to European problems, basically diseases such as TB, smallpox, and syphilis. They believed their sanitation problems could be solved using the same means as those in main European research centers: identification of the specific microbes, attenuation for the production of vaccines, sanitation and disinfection.

To understand the relation between microbiology and tropical medicine, we have to analyze the role played by the *Instituts Pasteur d'Oultre-mer* and mark the differences with Manson's works in *The London School of Tropical Medicine,* who, for the first time, made evident the peculiarities of that discipline. In the case of the Pasteur Institute, the fact of the tropical diseases having stopped white Europeans from establishing themselves in the colonies fostered the creation, after 1894 (the first institute was in Argelia), of a series of Overseas Institutes.

The mission of these Institutes was quite clear: to "export" the knowledge of the metropolitan laboratories, to create bacteriology laboratories, and to "qualify a new generation of autochthonous bacteriologists"(Bellora 1972, p. 283). The first evidence of this meeting between bacteriology and the tropics, mediated by military doctors, can be summarized on Dozon's statement: "While this conjunction became more precise (...), innumerous diseases, particularly the sleeping sickness, did not let themselves be reduced to experimental protocols nor to Pasteurian ideas" (Dozon 1991, p. 271). Many of these diseases offered resistance to both the specification of the causal agent and to the production of vaccines and sera, resistance that exist still today for Chagas' disease or sleeping sickness when we consider the lack of vaccines.

As Michel Morange stated: "The first obstacles for overseas Pasteurians was one of cultural order: it was a question of understanding the methods applied in France were not adequate for application in other countries" (Morange 1991, p. 240). It seems the tropics had much to learn (techniques, procedures, protocols), and the Pasteurians had little or nothing to teach.

This difficulty can be explained by the profound "alterity" associated to the tropics, and by the idea that this alterity should be modeled to the image of the "metropolis". This same difficulty was seen to exist in Argentina. It was then decided to place the threat from the tropics not overseas but on the opposite bank of the river, on the paludal jungle represented by the neighbor country, Brazil. A decision was made to reduce the variety of climates existing in Argentina to the moderate climate of Buenos Aires, and to identify the sanitary problems of this country to those of a Europe threatened by diseases which were not tropical.

Within this context, the fear of paludism and yellow fever occupied the same space as the European fears of the plagues from Africa and Asia: the threat posed by what is different. However, the climate in the north of Argentina, and also the vegetation and the fauna, were very similar or even identical to those of the south of Brazil. It is true that little was known, in the year of 1895, about the role of vectors in the transmission of diseases. Nevertheless, the obstinate denial of the role played by the mosquito in yellow fever was kept intact in the mind of the Head of the National Department of Hygiene, José Penna, until after 1916. In that year, his book *El paludismo y su profilaxia en Argentina* will be published (in association with Antonio Barbieri, head of the Prophylaxis Department of Paludism). It could be said this delayed text inaugurated the concern shown by Argentinean researchers towards so-called tropical diseases; in it, references are found to Ross, Grassi, and Manson.

This study conducted by Penna and Barbieri presents different levels of discourse. A quick epidemiological study makes evident the seriousness of the problem: the provinces of Tucuman, Salta, and Jujuy were several times hit by paludism, as well as Catamarca (1878) and Santiago del Estero where, in the year of 1902, almost 70% of the population were somehow affected by malaria.

A quick sociological analysis makes evident a reiteration of the old association between physical and moral conditions:

> Paludism has, during long years, constituted a barrier to immigrant movements, discouraging the foreigner from entering these regions which have been painted by fame with sinister colors. During many generations malaria has set upon the faces of the inhabitants of those areas the characteristic mark of its chronic form, cutting down on physical activities, dulling intelligence, weakening the organism and rendering men indifferent and apathetic in the fight for life (Penna 1916, p. 35).

The prophylactic strategies, as proposed, were based on the preventive use of quinine (for those less than 8-years-old, two chocolates per day; those older than 8 received two sweet tablets, having three and five grams of quinine each, respectively); in second place, this dose combats mosquitoes and larvae.

For the first time a discussion will be conducted on the need to perform "studies and description of local insects"; due consideration given to the deficiencies existing in Argentina, a list is made of the Brazilian entomologist doctors: Lutz, Oswaldo Cruz, Chagas, and Fajardo. The Hygiene Department had hired the entomologist Arthur Neiva, from the Oswaldo Cruz Foundation, to work at the Argentinean National Institute of Bacteriology. In 1915, Neiva publishes his "Study of a few Argentinean anopheles and their relation with malaria" (Penna 1916, p. 42). It is thus inaugurated in Argentina this new field of studies by a Brazilian researcher.

Nevertheless, in relation to yellow fever, Penna will renew, still in 1916, the same position he will fight for during the Second Latin American Congress of Medicine of 1904 and also in his clinical observations of 1912: "The opinions on the etiology and pathogeny of yellow fever are not demonstrated; I insist on the belief that these facts need demonstration" (Penna 1916, p. 224).

Finally, we should say that although Brazilian and Argentinean hygiene were the direct heirs of Pasteurian programs and principles, Argentina disregards the problems introduced by tropical diseases and insists in reducing all its problems to those capable of being thought of within the parameters of Pasteurian research protocols. As for Brazil, it has to face its sanitary problems for which no answer could be found in this Pasteurian timetable, posing new questions and a new research program which integrates and synthesizes various studies: bacteriology, parasitology, and medical entomology. Thus, indifferent of a documentary verification of any historical connection between Brazilian researchers and the tradition of Mansonian tropical medicine, it seems both groups share one same research program where studies of microbiology, parasitology and medical entomology could be complementary and unified.

It cannot be denied that, for Manson, the studies of entomology, zoology and studies he calls "of the naturalists", are not an accessory or secondary element, not just a complement for bacteriologic studies conducted in the laboratory, but rather a constitutive disciplinary space, like microbiology, of tropical medicine. "The medicine student (particularly that of tropical medicine) must be a naturalist before he can become an epidemiologist, a pathologist, or a doctor capable of working in his/her practice" (Manson 1984, p. xvi). Knowledge and identification of the infinite variety of tropical flora and fauna might contribute to reveal the mysteries of these diseases for which the causes remain unknown. This requires a detailed cognition of Brazilian local species, as well as a great variety of microbiology studies, to establish the relation between these local species, the known infectious diseases, and other region-specific diseases such as Chagas'.

These studies did not bring as the only result knowledge of the diversity and natural richness of the tropics: they made also possible to see the evidence of relinquishment, poverty, and the misery in the inner lands of Brazil. Those researchers who organized medical expeditions to the interior of Brazil were aware that to fight this reality it would be necessary to have knowledge of the particularities and the diversity of circumstances, both natural and social, which, conjugated, caused the epidemics.

In the words of Anne Marie Moulin:

> Tropical medicine on the days of Pasteur was connected with two important scientific contexts. On the one hand, the model of the microbiology laboratory (virus attenuation, experimental studies with animals); on the other hand, the field studies of parasitology, dominated by the study of disease-transmitting vectors and by the notion of natural cycles which suggested the need to dissect the complex environmental (ecological) interactions. The Pasteurian agenda and its triumphal spirit favor microbiology more than parasitology, action

upon germs, and the human reservoir upon a global perspective (Moulin 1996, p. 174)

References

Arnold, D. (1996), *Warm Climates and Western Medicine: The Emergence of Tropical Medicine*, Atlanta: Rodopi.

Bellora, A. (1972), *La salud pública*, Buenos Aires: Centro Editor de América Latina.

Benchimol, J. (1999), *Dos Micróbios aos Mosquitos*, Rio de Janeiro: Editora Fiocruz/UFRJ.

Canguilhem, G. (1989), "Prefacio", in Delaporte, F., *Historia de la fiebre amarilla*, México: CEMCA-UNA.

Chagas, C. (1981), *Coletánea de Trabalhos Científicos*, Brasilia: Universidade de Brasilia.

Coni, E. (1918), *Memorias de un médico higienista*, Buenos Aires: Biblioteca Médica Argentina.

Cruz, O. (1894), "Contribuição ao estudo da microbiologia tropical", *Brazil-Medico* 8 (37): 292-293.

Cruz, O. (1901), "Entomología, Contribuição para o estudo dos culicidios de Rio de Janeiro", *Brazil-Medico* 15 (43): 423-426.

Cruz, O. (1906), "Entomología: un nuevo género da sub-familia Anofelina", *Brazil-Medico* 15 (43): 423-426.

Cruz, O. (1910), "Prophylaxis of Malaria in Central and Southern Brazil", in Ross, R. (ed.), *The Prevention of Malaria*, London: John Murray, 1910.

Darmon, P. (1999), *L'Homme et les microbes*, Paris: Fayard.

Dozon, J.-P. (1991), "Pasteurisme, médecine militar et colonisation de Afrique noire", in Morange, M. (ed.), *L'Institut Pasteur: contributions à son histoire*, Paris: La découverte, 1991, pp. 269-278.

Löwy, I. (1991), "La mission del Institut Pasteur à Rio de Janeiro: 1901-1905", in Morange, M. (ed.), *L'Institut Pasteur: contributions à son histoire*, Paris: Ed. La decouverte, 1991, pp. 279- 295.

Machado, R. *et al.* (1978), *Danação da Norma*, Rio de Janeiro: Graal.

Manson, P. (1898), *Tropical Diseases*, London: Cassell and Company.

Moulin, A.M. (1996), "Tropical without the Tropics: The Turning-point of Pastorian Medicine in North Africa", in Arnold, D. (ed.), *Warm Climates and Western Medicine: the Emergence of Tropical Medicine*, Atlanta: Rodopi, 1996, pp. 160-180.

Michel, M. & J.P. Bado (1991), "Sur les traces du docteur Émile Marchoux: pionner de l'Institut Pasteur en Afrique noire", in Morange, M. (ed.), *L'Institut Pasteur: contributions à son histoire*, Paris: Ed. La decouverte, 1991, pp. 296-311.

Peard, J. (1996), "Tropical Medicine in Nineteenth-Century Brazil: The Case of the 'Escola Tropicalista Bahiana', 1860-1890", in David, A. (ed.), *Warm Climates and Western Medicine: The emergence of Tropical Medicine*, Atlanta: Rodopi, 1996, pp. 108-130.

Power, H. & L. Wilkinson (1998), "The London and Liverpool Schools of Tropical Medecine 1898-1998", in David Warrell (ed.), *Tropical Medicine: Achievements and Prospects*, London: The Royal Society of Medicine Press, 1998, pp. 281-292

Rawson, G. (1891), *Escritos y discursos* (coleccionados por A. Martínez), 2 Vols., Buenos Aires: Ed. Ceylan.

Penna, J. (1904), "El microbio y el mosquito en la patogenia y transmisión de la Fiebre Amarilla", in *Anales del II Congreso Medico Latino-Americano*, Buenos Aires, pp. 277-327.

Penna, J. & A. Barbieri (1916), *El Paludismo y su profilaxis en Aegentina*, Buenos Aires: Editora del Departamento Nacional de Higiene.

Stepan, N. (1976), *Gênese e Evolução da ciência Brasileira*, Rio de Janeiro: Artenova.

Wilde, E. (1885), *Curso de higiene pública*, Buenos Aires: Biblioteca Médica Argentina.

Primary sources

Anales del II Congreso Medico Latino Americano (1904). Buenos Aires. Sesiones del día 8 y 9 de abril.

Anales del III Congreso Médico Latino Americano (1907). Montevideo.

Anales del V Congreso Medico Latino-Americano (1913). Lima.

www.ingramcontent.com/pod-product-compliance
Lightning Source LLC
Chambersburg PA
CBHW050123170426
43197CB00011B/1696